Barriers and O
2-Year and 4-Year STEM Degrees

SYSTEMIC CHANGE TO SUPPORT STUDENTS' DIVERSE PATHWAYS

Committee on Barriers and Opportunities in
Completing 2-Year and 4-Year STEM Degrees

Shirley Malcom and Michael Feder, *Editors*

Board on Science Education
Division of Behavioral and Social Sciences and Education

Board on Higher Education and the Workforce
Policy and Global Affairs

National Academy of Engineering

The National Academies of
SCIENCES · ENGINEERING · MEDICINE

THE NATIONAL ACADEMIES PRESS
Washington, DC
www.nap.edu

THE NATIONAL ACADEMIES PRESS   500 Fifth Street, NW   Washington, DC 20001

This study was supported by Contract No. HRD-1244829 from the National Science Foundation, Contract No. B2012-30 from the Alfred P. Sloan Foundation, and an unnumbered contract from S.D. Bechtel, Jr. Foundation, with additional support from the National Academy of Sciences Kellogg Fund. Any opinions, findings, conclusions, or recommendations expressed in this publication do not necessarily reflect the views of any organization or agency that provided support for the project.

International Standard Book Number-13:   978-0-309-37357-9
International Standard Book Number-10:   0-309-37357-3
Library of Congress Control Number: 2016939799
Digital Object Identifier: 10.17226/21739

Additional copies of this report are available from the National Academies Press, 500 Fifth Street, NW, Keck 360, Washington, DC 20001; (800) 624-6242 or (202) 334-3313; http://www.nap.edu.

Cover credit: iStock image #34464984, ©kali9.

Suggested citation: National Academies of Sciences, Engineering, and Medicine. (2016). *Barriers and Opportunities for 2-Year and 4-Year STEM Degrees: Systemic Change to Support Diverse Student Pathways*. Committee on Barriers and Opportunities in Completing 2-Year and 4-Year STEM Degrees. S. Malcom and M. Feder, Editors. Board on Science Education, Division of Behavioral and Social Sciences and Education. Board on Higher Education and the Workforce, Policy and Global Affairs. Washington, DC: The National Academies Press. doi: 10.17226/21739.

The National Academies of
SCIENCES · ENGINEERING · MEDICINE

Preface

Policy makers agree that the nation's economic and social development require investment in the education of everyone. The level of that education and the skills required in 21st century America differ widely from those needed in the country inhabited and built by our forebears. The pace of change is different, as are the demographics of the U.S. population. While education in general is critical to the nation's future, it is widely recognized that the specific skills often acquired in the study of science, technology, engineering, and mathematics (STEM) fields are increasingly needed across the economy, and it is those fields that we have explored in depth in this report.

The decision to focus on specific fields was partly based on practical considerations. The scope of the study needed to be bounded so that a detailed report could be produced, and the national focus on STEM education and jobs led to the need to clarify what research can contribute to the ongoing policy debates. However, while the committee acknowledges the importance of STEM to the nation's economic competitiveness, we also recognize the importance of the pursuit of all knowledge, including the arts and humanities, and how these non-STEM areas also support the growth of ideas and solutions needed to address global challenges.

We also recognize that those holding STEM degrees have higher salaries and lower levels of unemployment, and there is a smaller pay gap between men and women in many STEM fields than in other fields. At the same time, we note that most people with STEM degrees are not working in STEM fields.

We do not tie our discussion to questions of the adequacy, oversupply, or surfeit of STEM degree holders. We note that those with an interest should be afforded an opportunity for success. STEM degrees not only provide credentials that attest to mastery of knowledge in specific STEM fields, but also indicate that the individuals likely possess skills that are used and valued in a variety of sectors of the economy. Beyond the interest in providing knowledge and skills that will be valuable in the economy is the value of having such knowledge and skills to support responsible citizenship in a pluralistic democracy. Study of STEM fields can enrich individuals as they engage in multiple roles across society.

Our forebears lived in a time when there were different norms as to the role of women and minorities in the community and the economy. Today, women are the majority of students in higher education. The shifting demographic means that the nation has to develop talent from across society, including among those who may not in the past have been afforded a quality education or those for whom society has not had expectations for success in STEM fields.

As we have explored the research to inform the question of STEM degree completion, we have tried to look to the extent possible at various groups in the population, especially at groups who, history shows, may not have been enabled to contribute to the talent pool for STEM. We know, for example, that in addition to women and underrepresented minorities, persons with disabilities and first-generation college students have faced barriers. Unfortunately, we have not always had robust data or relevant research to be able to outline the nature of those barriers or the opportunities to address them. To respond to this lack of guidance, we can only advocate that reforms be learner centered and that the system be viewed from the perspective of the learners.

Shirley Malcom, *Chair*
Committee on Barriers and Opportunities in
Completing 2-Year and 4-Year STEM Degrees

Acknowledgments

This report represents the work of thousands of individuals, not only those who served on the committee, wrote papers for it, and participated in the committee's open sessions, but also those who conducted and were the subjects of the research on which our conclusions and recommendations are based. We recognize their invaluable contributions to our work. The first thanks are to my fellow committee members, for their deep knowledge and contributions to the study.

This report was made possible by the important contributions of the National Academies of Sciences, Engineering, and Medicine, the study committee, and many other experts. We acknowledge the sponsorship of the Alfred P. Sloan Foundation, the S.D. Bechtel, Jr. Foundation, and the National Science Foundation (NSF). We particularly thank Elizabeth Boylan (program director, Alfred P. Sloan Foundation), Lisa Lomenzo (program officer, S.D. Bechtel, Jr. Foundation), Susan Singer (director, NSF Division of Undergraduate Education), and John Rand (program director, NSF Human Resources Development Division).

Over the course of this study, members of the committee benefited from discussion and presentations by the many individuals who participated in our three fact-finding meetings. At the first meeting, the committee discussed the charge with representatives from the Alfred P. Sloan Foundation, S.D. Bechtel, Jr. Foundation, and NSF. Carl Wieman (Stanford University) and Linda Slakey (independent consultant) discussed the major issues in science, technology, engineering, and mathematics (STEM) education to which the committee should attend. Lynne Molter (Swarthmore College) and Ann-Barrie Hunter (University of Colorado) described what can be derived from existing data on students seeking STEM degrees. Charles Henderson (Western Michigan University) presented evidence on strategies for creating and implementing changes in undergraduate STEM education. Lindsey Malcom-Piqueux (George Washington University) described the role of minority-serving institutions in educating students who aspire to earn a STEM degree.

During the second meeting, the committee heard expert testimony on the state of reform efforts in mathematics education from David Bressoud (Macalester College), Lou Gross (University of Tennessee), and Steven Ritter (Carnegie Learning). Casey George-Jackson (University of Illinois at Urbana–Champaign) and Rita Kirshtein (American Institute of Research) discussed the cost and price of STEM degrees. Evidence of the impact of authentic STEM experiences for students was presented by Elizabeth Ambos (Council on Undergraduate Research), Phillip Bowman (University of Michigan), and Kevin Egan (University of California, Los Angeles). The value of taking a systems approach to improving undergraduate STEM education was discussed by Ann Austin (Michigan State University), Ellen Goldey (Wofford College), Robert Hilborn (American Association of Physics Teachers), and Karan Watson (Texas A&M University).

The third committee meeting was structured as a public workshop on undergraduate STEM education. The workshop included two panel discussions on the goals and processes for reforming undergraduate STEM education. The first panel included Ryan Kelsey (Helmsley Trust), Mary Beth Oyer (Lockheed Martin), Dale Ramenzani (Corporate and University Relations Group), and Mercedes Talley (W.M. Keck Foundation). The second panel included Steve Barkanic (Business-Higher Education Forum), Emily Miller (Association of American Universities), Muriel Poston (Project Kaleidoscope Advisory Board), and Kacy Redd (Association of Public and Land-grant Universities). The meeting also included expert presentations on student persistence in STEM degrees at different types of institutions (2-year, 4-year, public, private, nonprofit, for profit, etc.) by Kevin Eagan (University of California, Los Angeles), Kevin Kinser (University of Albany, State University of New York), and Marco Molinaro (University of California, Davis). Mica Estrada (California State University, San Marcos) and Kim Godsoe (Brandeis University) discussed the impact of co-curricular supports for STEM students. Rafael Alvarez (San Diego City College), Cher Carrera (Santa Ana College), Mark Filowitz (California State University, Fullerton), Benjamin Flores (University of Texas at El Paso), and John Matsui (University of California, Berkeley) provided an overview of the programs that improve STEM student retention and persistence. The barriers and opportunities created by articulation agreements and transfer policies were discussed by Elizabeth Bejar (Florida International University), Randy Kimmens (Maricopa Community College), and Ken O'Donnell (California

State University). The committee also benefited from discussion by Noah Finkelstein (University of Colorado Boulder) and Omar Torres (College of the Canyons) about creating and sustaining systemic change.

We are grateful for the efforts of the 12 authors who prepared background papers:

- David Bressoud, on the barriers that STEM degree seekers encounter with mathematics
- Ken O'Donnell, on the regulations and policies affecting the transfer of credit between 2-year and 4-year institutions
- Mica Estrada, on the co-curricular supports for underrepresented students seeking a STEM degree
- Kevin Eagan, Tanya Figueroa, Brice Hughes, and Sylvia Hurtado, on the pathways to a STEM degree among students who begin college at a 4-year institution
- Michelle Van Noy and Matthew Zeidenberg, on the contributions of community colleges to undergraduate STEM education and workforce development
- Kevin Kinser, on the contributions of for-profit institutions to undergraduate STEM education and workforce development
- Hal Salzman and Michelle Van Noy, on STEM student pathways from 4-year institutions to 2-year institutions

This report has been reviewed in draft form by individuals chosen for their diverse perspectives and technical expertise, in accordance with procedures approved by the National Academies of Sciences, Engineering, and Medicine's Report Review Committee. The purpose of this independent review is to provide candid and critical comments that will assist the institution in making its published report as sound as possible and to ensure that the report meets institutional standards for objectivity, evidence, and responsiveness to the study charge. The review comments and draft manuscript remain confidential to protect the integrity of the deliberative process.

We thank the following individuals for their review of this report: Richard M. Amasino, Department of Biochemistry, University of Wisconsin–Madison; Eric Bettinger, Graduate School of Education, Stanford University; David Bressoud, Mathematics, Statistics, and Computer Science Department, Macalester College; Benjamin Flores, Department of Electrical and Computer Engineering, University of Texas at El Paso; Charles Henderson, Physics Department, Western Michigan University; Lynne Molter, Department of Engineering, Swarthmore College; Peter F. Murray, Science Learning Institute, Foothill College; Peter J. Polverini, School of Dentistry, University of Michigan; Linda L. Slakey, Dean Emerita, University of Massachusetts Amherst; Candice Thille, Graduate School of

Education, Stanford University; Diane C. Tucker, Science and Technology Honors Program, Honors College, University of Alabama at Birmingham; Jessica Utts, Department of Statistics, University of California, Irvine; and Carl Wieman, Department of Physics and Graduate School of Education, Stanford University.

Although the reviewers listed above provided many constructive comments and suggestions, they were not asked to endorse the content of the report nor did they see the final draft of the report before its release. The review of this report was overseen by Paul R. Gray, executive vice chancellor and provost emeritus, University of California, and Ron S. Brookmeyer, Department of Biostatistics and The Fielding School of Public Health, University of California, Los Angeles. Appointed by the Academies, they were responsible for making certain that an independent examination of this report was carried out in accordance with institutional procedures and that all review comments were carefully considered. Responsibility for the final content of this report rests entirely with the author and the institution.

Thanks are also due to the project staff. Michael Feder, of the Academies Board on Science Education, directed the study and played a key role in the report drafting process. Kelly Arrington managed the study's logistical and administrative needs, making sure meetings and workshops ran efficiently and smoothly. Joanna Roberts and Miriam Scheiber managed the manuscript preparation. Melissa Welch-Ross (senior program officer with the Institute of Medicine) summarized the research on the cost and price of STEM degrees. We are also grateful to Argenta Price (Christine Mirzayan science and technology fellow) for her synthesis of the research on undergraduate research experiences. Staff of the Division of Behavioral and Social Sciences and Education also provided help: Eugenia Grohman substantially improved the readability of the report; Kirsten Sampson Snyder expertly guided the report through the Academies report review process; and Yvonne Wise masterfully guided the report through production. Catherine "Kitty" Didion (senior program officer with the National Academy of Engineering), Jay Labov (senior advisor for education and communication), and Elizabeth O'Hare (program officer with the Board on Higher Education and the Workforce) provided critical guidance and input at each stage of the study. Finally, this study would not have been possible without the vision of Margaret Hilton (senior program officer with the Board on Science Education).

Shirley Malcom, *Chair*
Committee on Barriers and Opportunities in
Completing 2-Year and 4-Year STEM Degrees

Contents

Appendixes

Summary

Why do many of the students who enter higher education with an interest in pursuing study in science, technology, engineering, and mathematics (STEM) lose that interest before degree completion? How can the quality of the educational experience of undergraduate STEM students be improved? Motivated by these questions, the National Academies of Sciences, Engineering, and Medicine appointed the Committee on Barriers and Opportunities in Completing 2-Year and 4-Year STEM Degrees to address the barriers that prevent students from earning the STEM degrees to which they aspire and to identify opportunities to promote completion of undergraduate STEM degrees.

The committee approached its review of research on undergraduate STEM education from the viewpoint that all students who are interested in a STEM credential should be enabled to make an informed decision about whether a STEM degree is the right degree choice for them; afforded the opportunity to earn the degrees they seek with a minimum of obstacles; and supported by faculty, advisers, mentors, and institutional policies rather than being or perceiving themselves as being pushed out of STEM majors.

A diverse range of students take varied paths to earn STEM degrees. There are both differences and similarities across disciplines, institution types, and student characteristics. Contrary to the image of a linear route to a bachelor's degree in STEM (often referred to as the STEM pipeline), we found instead a complex array of pathways to a varied set of undergraduate credential outcomes, both 2-year and 4-year degrees. Students use 2-year and 4-year institutions in ways likely not envisioned by educators and policy makers, with frequent transfers, concurrent enrollment at multiple institutions, and multiple points of entry, exit, and reentry to the pathways.

Such pathways have major implications for the financing of, the time to, and the cost of degrees. However, existing data systems make it difficult to track students seeking STEM degrees because they focus on first-time, full-time students; such students account for a minority of the undergraduate population. And the diversity of pathways, even for those who may successfully complete STEM degrees, raises serious practical questions about the validity of the accountability metrics being used or proposed for higher education institutions.

The very culture of STEM presents both barriers and opportunities for successful degree completion for all students. The normative culture of STEM can be a barrier for students from underrepresented groups because it often includes views of student ability as inherent or natural, related to one's genetics, and thus not amenable to improvement. Related to this view is the tendency for introductory mathematics and science courses to be used as "gatekeeper" courses with highly competitive classroom environments that may discourage students who are new to the fields, especially women and those from minority backgrounds.

Institutional, state, and national education policies have not been developed to support the various pathways that students are now taking to earn a STEM degree. Transfer and articulation policies (or the lack of these) often slow students' progress to degrees, deter students from transferring, and increase the cost of their undergraduate education. In addition, students often pay more for a STEM degree than expected due to tight course sequencing, degree requirements, grading policies, the need for developmental coursework, and the availability of courses. The high cost of providing some STEM degrees and diminishing funding from state and federal sources have led some universities to adopt the practice of charging differential tuition. While research on the effects of differential pricing is limited, existing studies indicate potentially negative effects of this policy on selecting a STEM major, particularly among women and underrepresented minorities.

Some states have adopted performance-based funding formulas, which reward institutions with higher graduation rates. This policy is feared to have the unintended consequence of placing a greater focus on graduation rates rather than either the quality of the degrees offered or on the populations being served, but studies have yet to explore whether these fears are justified. It also has been criticized for failing to recognize the work being done by institutions that are attempting to support STEM degree completion by capable students who come from different profiles—such as those who are academically less well prepared, including many from underrepresented groups. The policy of performance funding may also have had the unintended consequence of limiting the recruitment and enrollment of students from those groups, who may be deemed at high risk of failure, both generally and in STEM fields.

Some colleges that provide co-curricular support to students (such as peer tutoring, research experiences, and living-learning communities) and have improved instructional strategies have seen improvement in student outcomes. These structures often function outside of the regular operations of the departments. However, an institution-wide or systemic approach to change is most likely to yield meaningful and lasting results.

Overall, it is clear that the STEM pipeline metaphor is not an accurate portrayal of the diverse, complex paths that students take to earn STEM degrees. The prominent practice of undertaking piecemeal reform efforts has typically been shown to be unsuccessful because these efforts do not attend to complex pathways being taken to earn STEM degrees, the challenges the students face along those pathways, and the policy environments in which these challenges are addressed. To address the needs of STEM students, colleges, universities, federal agencies, professional organizations, state and federal policy makers, accrediting agencies, foundations, and STEM departments need to work together, across their individual structures, to create comprehensive and lasting improvements to undergraduate STEM education.

CONCLUSIONS

CONCLUSION 1 There is an opportunity to expand and diversify the nation's science, technology, engineering, and mathematics (STEM) workforce and STEM-skilled workers in all fields if there is a commitment to appropriately support students through degree completion and provide more opportunities to engage in high-quality STEM learning and experiences.

CONCLUSION 2 Science, technology, engineering, and mathematics (STEM) aspirants increasingly navigate the undergraduate education system in new and complex ways. It takes students longer for completion of degrees, there are many patterns of student mobility within and across institutions, and the accommodation and management of student enrollment patterns can affect how quickly and even whether a student earns a STEM degree.

CONCLUSION 3 National, state, and institutional undergraduate data systems often are not structured to gather information needed to understand how well the undergraduate education system and institutions of higher education are serving students.

CONCLUSION 4 Better alignment of science, technology, engineering, and mathematics (STEM) programs, instructional practices, and student supports is needed in institutions to meet the needs of the populations they serve. Programming and policies that address the climate of STEM departments and classrooms, the availability of instructional supports and authentic STEM experiences, and the implementation of effective teaching practices together can help students overcome key barriers to earning a STEM degree, including the time to degree and the price of a STEM degree.

CONCLUSION 5 There is no single approach that will improve the educational outcomes of all science, technology, engineering, and mathematics (STEM) aspirants. The nature of U.S. undergraduate STEM education will require a series of interconnected and evidence-based approaches to create systemic organizational change for student success.

CONCLUSION 6 Improving undergraduate science, technology, engineering, and mathematics education for all students will require a more systemic approach to change that includes use of evidence to support institutional decisions, learning communities and faculty development networks, and partnerships across the education system.

RECOMMENDATIONS

RECOMMENDATION 1 Data collection systems should be adjusted to collect information to help departments and institutions better understand the nature of the student populations they serve and the pathways these students take to complete science, technology, engineering, and mathematics degrees.

RECOMMENDATION 2 Federal agencies, foundations, and other entities that fund research in undergraduate science, technology, engineering, and mathematics (STEM) education should prioritize research to assess whether enrollment mobility in STEM is a response to financial, institutional, individual, or other factors, both individually and collectively, and to improve understanding of how student progress in STEM, in comparison with other disciplines, is affected by enrollment mobility.

RECOMMENDATION 3 Federal agencies, foundations, and other entities that support research in undergraduate science, technology, engineering, and mathematics education should support studies with multiple methodologies and approaches to better understand the effectiveness of various co-curricular programs.

RECOMMENDATION 4 Institutions, states, and federal policy makers should better align educational policies with the range of education goals of students enrolled in 2-year and 4-year institutions. Policies should account for the fact that many students take more than 6 years to graduate and should reward 2-year and 4-year institutions for their contributions to the educational success of students they serve, which includes not only those who graduate.

RECOMMENDATION 5 Institutions of higher education, disciplinary societies, foundations, and federal agencies that fund undergraduate education should focus their efforts in a coordinated manner on critical issues to support science, technology, engineering, and mathematics (STEM) strategies, programs, and policies that can improve STEM instruction.

RECOMMENDATION 6 Accrediting agencies, states, and institutions should take steps to support increased alignment of policies that can improve the transfer process for students.

RECOMMENDATION 7 State and federal agencies and accrediting bodies together should explore the efficacy and tradeoffs of different articulation agreements and transfer policies.

RECOMMENDATION 8 Institutions should consider how expanded and improved co-curricular supports for science, technology, engineering, and mathematics (STEM) students can be informed by and integrated into work on more systemic reforms in undergraduate STEM education to more equitably serve their student populations.

RECOMMENDATION 9 Disciplinary departments, institutions, university associations, disciplinary societies, federal agencies, and accrediting bodies should work together to support systemic and long-lasting changes to undergraduate science, technology, engineering, and mathematics education.

1

Introduction

Interest in science, technology, engineering, and mathematics (STEM) credentials continues to grow among high school graudates who plan to attend a 2-year or 4-year institution (National Science Board, 2014; National Center for Education Statistics, 2013).[1] At the same time, calls for improvements to undergradute STEM education persist in part because the 6-year completion rates for STEM degrees remain around 40 percent (President's Council of Advisors on Science and Technology, 2012): this is noticeably lower than the rate of 56 percent among all students who first enrolled in 2007 in all types of 2-year and 4-year institutions (Shapiro et al., 2013). It is important to consider whether students interested in earning a STEM degree leave STEM for reasons related to how STEM is taught or the nature of the learning environments, in contrast to leaving STEM because they discover a different course of study that is a better match for their interests and abilities.

A recent report to the President, *Engage to Excel: Producing One Million Additional College Graduates with Degrees in Science, Technology, Engineering and Mathematics* (President's Council of Advisors on Science and Technology, 2012), cited the need to develop an adequate base of talent in STEM fields to ensure the economic strength, national security, global competitiveness, environment, and health of the United States. Industry and business leaders also have expressed concern about having adequate

---

[1]In this report, we use the term "institution" to refer to colleges and universities. We refer to 2-year institutions and community colleges interchangeably even though some community colleges grant 4-year degrees.

numbers of STEM graduates at the baccalaureate and associate levels. At the same time, a number of researchers have examined trends in the data and come to conflicting conclusions regarding whether there is or will be a shortage of graduates with STEM degrees (see, e.g., Carnevale et al., 2011; Rothwell, 2013; Salzman, 2013). Some analysts estimate a shortfall of STEM graduates in the next 10 years (Carnevale et al., 2011), while others suggest a surplus of STEM graduates over the same period of time (Salzman, 2013). Different conclusions seem to arise due to disagreements about a number of fundamental assumptions. For example, there is not agreement about what jobs should be included as part of the STEM workforce. Research on the current and future STEM workforce continues to attempt to resolve these contradictions among economic and workforce forecasts.

The heightened attention to workforce predictions has focused most of the attention on undergraduate STEM education reform on the question of workforce demand, rather than on whether institutions are providing students with a high-quality education and the supports they need to complete a STEM credential.[2] Our task was different: we do *not* consider questions of shortage, adequacy, or surfeit. Rather, as directed by the statement of task for the study (see Box 1-1), our work centered on the barriers and opportunities that students encounter along the increasingly diverse pathways to earning a STEM credential at a 2-year or 4-year institution. We thus have focused on research that investigates the roles that people, processes, and institutions play in 2-year and 4-year STEM credential production. We have done so with the view that all undergraduate students interested in a STEM credential should be

- enabled to make an informed decision about whether a STEM credential is the right choice for them;[3]
- afforded the opportunity to earn the credential they seek with a minimum of obstacles; and
- supported by faculty, advisers, mentors, and institutional policies rather than being or perceiving themselves as being pushed out of STEM majors or having to overcome what they perceive as insurmountable obstacles.

---

[2] A "credential" is any degree or certification that can be earned by a student at 2-year or 4-year institutions.

[3] By this, we mean to stress that it should be expected that some students who initially seek a STEM degree will choose a different discipline to major in because they find that they do not like the STEM discipline they were originally interested in or they find an alternate discipline that is a better match for their interests and abilities. Such choices should be viewed as a positive outcome, because it is part of the natural process of exploration and discovery in college. On the other hand, it would be a major concern if students choose a non-STEM major because they have negative experiences in STEM programs for which they are otherwise a good match.

BOX 1-1
Statement of Task

An ad hoc committee will conduct a comprehensive study to understand the barriers facing 2-year and 4-year undergraduates who intend to major in science, technology, engineering, and mathematics (STEM) and opportunities for overcoming these barriers. The committee will prepare a report that will include conclusions based on the evidence and provide research-based guidance to inform policies and programs that aim to attract and retain students to complete associate's and bachelor's degrees in STEM disciplines.

This report gives special attention to factors that influence diverse students' (e.g., by race, ethnicity, gender, and socioeconomic factors) decisions to enter into, stay, or leave majors in STEM fields. We explore factors inclusive of and beyond the quality of instruction, such as grading policies, course sequences, undergraduate learning environments, student supports, co-curricular activities, students' general self-efficacy and self-efficacy in science, family background, and governmental and institutional policies that affect STEM educational pathways. The report explores the role of motivation, interest, and attitude in shaping undergraduates' trajectories in STEM, especially in the transition from 2-year to 4-year institutions.

This study builds on previous work of the National Academies of Sciences, Engineering, and Medicine, including the reports *Community Colleges in the Evolving STEM Education Landscape* (National Research Council and National Academy of Engineering, 2012), *Expanding Underrepresented Minority Participation: America's Science and Technology Talent at the Crossroads* (National Research Council, 2011), *Discipline-Based Education Research: Understanding and Improving Learning in Undergraduate Science and Mathematics* (National Research Council, 2012), *The Engineer of 2020: Visions of Engineering in the New Century* (National Academy of Engineering, 2004), and *Changing the Conversation: Messages for Improving Public Understanding of Engineering* (National Academy of Engineering, 2008).

WHAT WE MEAN BY STEM

Any thoughtful discussion of STEM education requires a working definition of what constitutes STEM disciplines. While STEM is a term commonly used, an enduring question for policy makers, advocates, researchers, and this committee is what fields of study and practice are included in STEM. Despite legal definitions and the policies based on them, there still is little consensus as to which fields and courses of study should fall within STEM.

STEM has been previously defined by the National Academy of Engineering and National Research Council (2009, p. 17):

- Science is the study of the natural world, human behavior, interaction, and social and economic systems. It includes studies of the laws of nature associated with physics, chemistry, and biology and the treatment or application of facts, principles, concepts, or conventions associated with these disciplines.
- Technology comprises the entire system of people and organizations, knowledge, processes, and devices that go into creating and operating technological artifacts, as well as the artifacts themselves.
- Engineering is both a body of knowledge—about the design and creation of human-made products—and a process for solving problems. This process is design under constraint. One constraint in engineering design is the laws of nature, or science. Other constraints include factors such as time, money, available materials, ergonomics, environmental regulations, manufacturability, and reparability. Engineering utilizes concepts in science and mathematics as well as technological tools.
- Mathematics is the study of patterns and relationships among quantities, numbers, and shapes. Mathematics includes theoretical mathematics and applied mathematics.

The National Science Foundation (NSF) also delineates the STEM fields as physical, biological, earth, atmospheric and ocean sciences; mathematics, statistics, and computer sciences; social, behavioral, and economic sciences; and all areas of engineering and technology. In an examination of the research on STEM education, which covers an array of disciplines, the committee found that only some researchers used the NSF definition, while many studies did not include social and behavioral sciences. Inconsistencies in the definition of STEM can make it difficult to reconcile findings across studies. For this reason, throughout this report we note which fields are included in the STEM education research summarized.

Given that the focus of this report is to identify the barriers and opportunities to earning STEM degrees, we focused our review on STEM fields where attrition is most pronounced, particularly among underrepresented groups; is caused by similar barriers or factors (e.g., level of mathematics preparation and proficiency, departmental and classroom culture, course sequencing, and cost); and can be attenuated by similar interventions or systemic changes.

The committee identified some barriers and opportunities in completing a STEM degree that are common across STEM disciplines, and we found that some barriers and opportunities differ across them. Where relevant, we discuss the differences among STEM fields.

STEM DEGREE PATHWAYS

A frequent metaphor used to describe the movement of students toward STEM degrees is that of a pipeline, the implication being that they are on the road to a degree unless or until they "leak out." This metaphor does not begin to capture the complex ways that today's students use colleges and universities to complete their degrees. This report provides new ways of both envisioning and planning for the routes and strategies (or lack thereof) in and across institutions of higher education that today's students use in pursuit of STEM degrees.

In 2010, nearly 40 percent of entering students at 2-year and 4-year postsecondary institutions indicated an intention to major in STEM; an increase from 2007, when about 33 percent indicated the intention to major in STEM (National Science Board, 2014). Overall, numbers of STEM credentials are increasing for almost every STEM discipline. At the same time, about one-half of students with the intention to earn a STEM bachelor's degree and more than two–thirds of those intending to earn a STEM associate's degree fail to earn these degrees within 6 or 4 years, respectively (Eagan et al., 2014; Van Noy and Zeidenberg, 2014). In addition, many students who do complete credentials take longer than the advertised length of the programs (Eagan et al., 2014; Van Noy and Zeidenberg, 2014), for example, students aspiring to a B.S. in biology enter this course of study expecting to graduate in 4 years, based on the information provided to them by institutions and biology departments. The extended time to degree results in higher costs that students and their families may not have anticipated.

Understanding students' trajectories to STEM degrees and what causes them to stay or leave requires answers to a number of questions. Are the STEM educational pathways any less efficient than those for other fields of study? Are they more efficient for some students than for others? If so, what constitutes and contributes to effective patterns? At what points do losses occur? How might the losses be minimized and greater efficiencies realized? These questions are at the heart of the committee's study.

A better understanding of the current "system" of STEM degrees in 2-year and 4-year institutions has important implications for national education policy and planning. Efforts being undertaken by federal- and state-level agencies and departments and by private funders of higher education need to be informed by the best possible data and analysis about what works where, for whom, and under what circumstances. Much of the data that could help address these national priorities in education and workforce remain either uncollected, collected in idiosyncratic formats that make analyses difficult or impossible, or are mired in regulations that are rightly designed to protect student privacy but that hamper informed decision making at all levels of the education system. For example, it is difficult to track part-time students and students who transfer among institutions: both kinds of students are growing proportions of the overall undergraduate student population.

In order to tell this complex story, we have organized the report and our findings around the concept of pathways. In Chapter 2, we describe several broad pathways based on whether students first enroll in a 2-year or 4-year institution. We describe the pathways that community college students take when seeking to earn an associate's degree, to transfer to a 4-year institution (mostly, science and engineering majors), or to earn a certificate (mostly, technician majors). We also trace the pathways that students who first enroll at a 4-year institution take to earn a STEM degree, including whether they initially enter a STEM degree program or choose a STEM program later, and how students move across institutions. Those moves cover many combinations: transferring from a 2-year to a 4-year institution, reverse transferring from a 4-year to a 2-year institution, transferring between 2-year institutions, transferring between 4-year institutions, as well as combinations of attendance at multiple institutions.

We review what happens to those who do not complete the journey. We assess where students encounter barriers and how the barriers affect their education pathways. We describe the major changes in student demographics; how students view, value, and use programs of higher education; and how institutions can adapt to support successful student outcomes. In doing so, we question whether the definitions and characteristics of what constitutes success in STEM should change. As we explore these issues, we identify where further research is needed to build a system that works for all students who aspire to STEM credentials.

The questions and issues that we cover in this report are not all specific to STEM education. Some of the barriers and opportunities that we explore occur across all of undergraduate education. Thus, we draw from research from undergraduate education in general, as well as from STEM-specific education when possible. We also point out where trends or findings are applicable to both STEM and non-STEM pathways and which are unique to STEM.

THE NEW NORMAL IN UNDERGRADUATE STEM EDUCATION

Who are today's undergraduate students who aspire to earn STEM degrees? How do they compare with undergraduates more generally? What is known about those who switch out of STEM programs, those who are "undecided," or those who enter STEM after having first selected a different field of study, and those who leave higher education without completing any degree? Beyond interest and motivation, what prior preparation do STEM majors bring with them to college? What is it about their backgrounds and the culture and mission of the academic departments and institutions they enter that contribute to the current outcomes? Are some types of institutions and academic programs more or less successful in producing STEM graduates for different groups of students?

Answering these questions means probing deeply into the patterns of study for different groups of students. It also means throwing aside some of the misconceptions that persist about who is a STEM student. Historically, the conception of STEM undergraduates has been students fresh out of high school who enter a 4-year college and complete degrees in 4 years: this pattern has so changed that such students are less than half of the undergraduate population (Eagan et al., 2014; Salzman and Van Noy, 2014).

Undergraduate students pursue degrees in a wide range of types of institutions: research universities, comprehensive universities, and 2-year and 4-year colleges, as well as for-profit institutions. Community colleges play an increasingly important role in the national higher education system, including in STEM education (Mooney and Foley, 2011). In 2011, nearly half of all students at the undergraduate level attended 2-year colleges: the 7.5 million students in 2-year colleges were 42 percent of all undergraduates (National Center for Education Statistics, 2014). Associate's degrees comprised 33 percent of all undergraduate degrees awarded in 2008–2009 (National Center for Education Statistics, 2013).

The propensity to enroll in different types of institutions varies for different groups of students (Kena et al., 2014). Minority, first-generation, and low-income students disproportionately attend 2-year institutions. Fifty-seven percent of all black undergraduate students and 60 percent of all Hispanic undergraduate students attended community colleges in 2011–2012, compared with 41 percent of white and Asian/Pacific Islander undergraduate students (Witham et al., 2015). Students from families whose income is in the bottom or third quartile are 50 percent of the student body at 2-year institutions, but only 14–34 percent of the student body at competitive 4-year institutions (Witham et al., 2015). Enrollment patterns also differ by parental education level: 48 percent of undergraduate students whose parents did not complete high school attend a community college, 42 percent of students whose parents completed high school attend a community college, and 34 percent of students whose parents completed college attend a community college (Witham et al., 2015).

SUCCESS IN UNDERGRADUATE STEM EDUCATION

The most commonly used assessment of success in undergraduate education, the graduation rate,[4] is a popular metric for a number of reasons. Institutions, students, and policy makers like them because they perceive them to be aligned with the primary goal of most college students (Bailey and Xu, 2012). In addition, completion and progression data are widely available and can be more easily collected in a consistent manner than other outcomes, such as wages or employment.

Critics have pointed out, however, that graduation rates on their own are a flawed metric of success because they are influenced by factors beyond the control of an institution. Graduation rates also are influenced by the characteristics of the students who are accepted at each institution. Thus, highly selective institutions would be expected to have higher graduation rates than institutions that are less selective. In addition, a degree is not the ultimate goal of all college students, especially among students at 2-year institutions: they may also seek to transfer to 4-year institutions without earning a degree, to earn a certificate, or to learn job-related skills. Thus, graduation rates provide some indication of the success of an undergraduate STEM program, but this information is difficult to interpret without information regarding student preparation, student goals, and institutional context.

An even broader vision of success has been emerging from definitions of success developed by various stakeholder groups, including the American Association of Community Colleges, the Aspen Institute, the Bill & Melinda Gates Foundation, and the National Governors Association. These visions shift the focus to a broader set of academic indicators, such as success in remedial and first-year courses, course completion, credit accumulation, time to degree, retention and transfer rates, degrees awarded, student diversity, and learning outcomes. However, there are as yet no systemic, national data sources on such factors.

Specific frameworks for success have recently been developed by a number of groups. These frameworks include both academic indicators and factors associated with the quality of STEM education. For example, the Association of American Universities (AAU) framework for success in undergraduate STEM education focuses on improving undergraduate STEM instruction and the culture of the learning environments.[5] The framework includes three factors that need to be addressed together: pedagogy, scaffolding, and cultural change. Pedagogy includes aligning faculty incentives with high-quality instructional practices, leadership commitment

---

[4]Degree completion at 4-year institutions is typically based on a 6-year time frame, and a 4-year time frame is used for degrees and certifications at 2-year institutions.

[5]For more information, see https://stemedhub.org/groups/aau/framework [April 2015].

to improved pedagogy, and assessing teaching practices. Under scaffolding, the AAU framework focuses on improved facilities, integrating technology into classroom instruction, faculty professional development, and the use of data for continuous improvement. Culture change includes ensuring expanded access, articulated learning goals, leadership commitment to change, establishing metrics of effective teaching practices, and the alignment of incentives with high-quality teaching practices. AAU is working on collecting data on these factors.

Some people in the field have also begun to include interpersonal and psychological factors as components of student success. Schreiner and colleagues (2010) have begun to focus on three key areas that contribute to student success and persistence: academic engagement and determination, interpersonal relationships, and psychological well-being. They identify thriving as a desirable goal for students, by which they mean more than surviving and graduating. Thriving means that students are engaged in the learning process, investing effort to reach important educational goals, managing their time and commitments effectively, connected in healthy ways to other people, optimistic about their future, positive about their present choices, appreciative of differences in others, and committed to making a contribution to their community (Schreiner et al., 2009).

As we approached our charge, we took the view that success is achieved when all students who are interested in STEM majors

- are able to make informed decisions about the best course of study for them based on interests, motivation, and career aspirations;
- understand the variety of potential career pathways that come with STEM degrees;
- have a clear understanding of STEM content and practices;
- do not face unreasonable barriers along their pathways that discourage them or make progress impossible; and
- are aware of connections between STEM and societal issues and concerns.

COMMITTEE APPROACH AND THE REPORT

This study was designed to describe the status of knowledge about the barriers faced by students with an interest in earning a STEM degree or certificate and the opportunities and strategies to remove these barriers (see the statement of task in Box 1-1). The report includes an in-depth analysis of the students who seek STEM degrees, the pathways taken to STEM degrees, the barriers to earning STEM degrees, programs and policies that support the completion of STEM degrees, and the systemic reforms needed to improve undergraduate STEM education for all students.

With support from the S.D. Bechtel, Jr. Foundation, the Alfred P. Sloan Foundation, and NSF, the National Academies of Sciences, Engineering, and Medicine established the Committee on Barriers and Opportunities in Completing 2-Year and 4-Year STEM Degrees to undertake this study. Selected to reflect a diversity of perspectives and a broad range of expertise, the 18 committee members included experts in the sociology of education; the current STEM workforce; higher education policy, practice, and administration; data collection methodologies; longitudinal and career research; educational and career counseling; STEM education reform; and advanced technical education. In addition, the committee included balanced representation across the range of state-supported and private universities and colleges, special-focus institutions, and 2-year colleges (see the biographical sketches of members in Appendix B).

In addressing the statement of task (Box 1-1), the committee focused its attention on students who aspire to earn a STEM credential, with the understanding that students in other fields also take STEM courses. For example, introductory STEM courses are required as part of general education credit requirements for students who aspire to a degree in many non-STEM fields (e.g., health sciences, humanities) at the vast majority of 2-year and 4-year institutions. We anticipate that the changes recommended in this report could lead to positive effects for a much larger pool of students than are the primary focus of this study.

The committee conducted its work through an iterative process of gathering information, deliberating on it, identifying gaps and questions, gathering further information to fill these gaps, and holding further discussions or seeking expert guidance. In our search for relevant information, we held three public fact-finding meetings and reviewed published and unpublished research reports and evaluations. We also commissioned seven white papers on a wide range of topics:

1. Regulations and policies affecting the transfer of credit between 2-year and 4-year institutions, by Ken O'Donnell.
2. Co-curricular supports for underrepresented students seeking a STEM degree, by Mica Estrada.
3. Pathways to a STEM degree among students who begin college at a 4-year institution, by Kevin Eagan, Tanya Figueroa, Brice Hughes, and Sylvia Hurtado.
4. Contributions of community colleges to undergraduate STEM education and workforce development, by Michelle Van Noy and Matthew Zeidenberg.
5. Contributions of for-profit institutions to undergraduate STEM education and workforce development, by Kevin Kinser.

6. The effect of mathematics education on the trajectories of STEM students, by David Bressoud.
7. STEM student pathways from 4-year institutions and 2-year institutions, by Hal Salzman and Michelle Van Noy.[6]

The committee as a whole met in person four times. At the first meeting, the committee discussed the charge with representatives from the Alfred P. Sloan Foundation and NSF. The meeting also included presentations from experts on issues related to student completion and persistence in STEM majors; creating and implementing changes to improve student outcomes; discipline-specific barriers, opportunities, and reform efforts; and serving underrepresented groups at 2-year and 4-year institutions.

During its second meeting, the committee heard expert testimony on the state of reform efforts in mathematics education; the cost and price of STEM degrees; the importance of and barriers to authentic STEM experiences for students;[7] and the value of taking a systems approach to improving undergraduate STEM education. Both meetings included private discussion among the committee members, which allowed them the opportunity to debate the relevance of the findings presented.

The third committee meeting was structured as a public workshop on undergraduate STEM education. The workshop included two panel discussions on the goals and processes for reforming undergraduate STEM education. The first panel included representatives from foundations and industries, and the second panel included representatives from national associations. The meeting also included expert presentations on and discussions of student persistence in STEM degrees at different types of institutions (2-year, 4-year, public, private, nonprofit, for-profit, etc.); cultural barriers within STEM departments and classrooms; co-curricular supports; models of transfer and articulation agreements/systems; and sustaining systemic change. Prior to the start of the workshop, the committee met for half a day to discuss the report outline and potential conclusion and recommendation topics.

At the fourth committee meeting, we intensely analyzed the relevant evidence that had been uncovered and discussed our conclusions. We were particularly focused on identifying bodies of research that are characterized by systematic collection and interpretation of evidence and exploring the ways in which these research literatures connect to each other.

The report takes a student-focused approach to identifying the barriers and opportunities to earning 2-year and 4-year undergraduate STEM

---

[6] The public meeting agendas and white papers are available at http://sites.nationalacademies.org/DBASSE/BOSE/CurrentProjects/DBASSE_080405 [April 2016].

[7] See p. 90 for definition of and discussion of authentic STEM experiences.

degrees or certifications, and this theme is reflected in all chapters of the report. In Chapter 2, we describe the pathways students take to earn STEM degrees. The chapter also provides a detailed look at who STEM degree seekers are, what institutions they attend, and how they navigate the undergraduate STEM education pathways. Differences in student pathways, majors, and institution type are highlighted throughout. Chapter 3 describes the effect of the culture of STEM departments and classrooms on students interested in a STEM credential. Chapter 4 provides a synopsis of instructional, departmental-level, and institutional-level barriers to STEM degrees and certifications. In addition, we review the effects of the range of interventions developed to improve student outcomes.

In Chapter 5, we review the system-level and policy barriers and the steps that can be taken to remove them. In Chapter 6, we describe how to create systemic and lasting change. The final chapter contains our conclusions about the barriers and opportunities for 2-year and 4-year undergraduate STEM education and presents our recommendations to faculty, STEM departments, colleges and universities, professional societies, higher education organizations, state governments, and the federal government to improve STEM education for all students interested in STEM degrees.

REFERENCES

Bailey, T., and Xu, D. (2012). *Input-Adjusted Graduation Rates and College Accountability: What Is Known from Twenty Years of Research?* HCM Strategists LLC, Bill & Melinda Gates Foundation. Available: http://www.hcmstrategists.com/contextforsuccess/papers/LIT_REVIEW.pdf [April 2015].

Carnavale, A., Smith, N., and Melton, M. (2011). *STEM: Science, Technology, Engineering, and Mathematics.* Georgetown University Center on Education and the Workforce. Available: http://files.eric.ed.gov/fulltext/ED525297.pdf [April 2015].

Eagan, K., Hurtado, H, Figueroa, T., and Hughes, B. (2014). *Examining STEM Pathways among Students Who Begin College at Four-Year Institutions.* Commissioned paper prepared for the Committee on Barriers and Opportunities in Completing 2- and 4-Year STEM Degrees, National Academy of Sciences, Washington, DC. Available: http://sites.nationalacademies.org/cs/groups/dbassesite/documents/webpage/dbasse_088834.pdf [April 2015].

Kena, G., Aud, S., Johnson, F., Wang, X., Zhang, J., Rathbun, A., Wildinson-Fliker, S., and Kristapovich, P. (2014). *The Condition of Education.* Washington, DC: U.S. Department of Education, National Center for Education Statistics.

Mooney, G.M., and Foley, D.J. (2011). *Community Colleges: Playing an Important Role in the Education of Science, Engineering, and Health Graduates.* Arlington, VA: National Science Foundation.

National Academy of Engineering. (2004). *The Engineer of 2020: Visions of Engineering in the New Century.* Washington, DC: The National Academies Press.

National Academy of Engineering. (2008). *Changing the Conversation: Messages for Improving Public Understanding of Engineering.* Committee on Public Understanding of Engineering Messages. Washington, DC: The National Academies Press.

National Academy of Engineering and National Research Council. (2009). *Engineering in K–12 Education: Understanding the Status and Improving the Prospects.* Committee on K–12 Engineering Education. L. Katehi, G. Pearson, and M. Feder (Eds.). Washington, DC: The National Academies Press.

National Center for Education Statistics (2013). *Digest of Education Statistics 2013.* Washington, DC: U.S. Department of Education.

National Center for Education Statistics. (2014). *Digest of Education Statistics 2014.* Washington, DC: U.S. Department of Education.

National Research Council. (2009). *Learning Science in Informal Environments: People, Places, and Pursuits.* Committee on Learning Science in Informal Environments, P. Bell, B. Lewenstein, A.W. Shouse, and M.A. Feder (Eds.). Board on Science Education, Center for Education. Division of Behavioral and Social Sciences and Education. Washington, DC: The National Academies Press.

National Research Council. (2011). *Expanding Underrepresented Minority Participation: America's Science and Technology Talent at the Crossroads.* Committee on Underrepresented Groups and Expansion of the Science and Engineering Workforce Pipeline. F.A. Hrabowski, P.H. Henderson, and E. Psalmonds (Eds.). Board on Higher Education and the Workforce, Division on Policy and Global Affairs. Washington, DC: The National Academies Press.

National Research Council. (2012). *Discipline-Based Education Research: Understanding and Improving Learning in Undergraduate Science and Engineering.* Committee on the Status, Contributions, and Future Directions of Discipline-Based Education Research. S. Singer, N.R. Nielsen, and H.A. Schweingruber (Eds). Board on Science Education, Division of Behavioral and Social Sciences and Education. Washington DC: The National Academies Press.

National Research Council and National Academy of Engineering. (2012). *Community Colleges in the Evolving STEM Education Landscape: Summary of a Summit.* S. Olson and J.B. Labov, Rapporteurs. Planning Committee on Evolving Relationships and Dynamics Between Two- and Four-Year Colleges, and Universities. Board on Higher Education and Workforce, Division of Policy and Global Affairs. Board on Life Sciences, Division on Earth and Life Sciences. Board on Science Education, Teacher Advisory Council, Division of Behavioral and Social Sciences and Education. Engineering Education Program Office, National Academy of Engineering. Washington, DC: The National Academies Press.

National Science Board. (2014). *Science and Engineering Indicators 2014.* Arlington VA: National Science Foundation (NSB 14-01).

President's Council of Advisors on Science and Technology. (2012). *Report to the President. Engage to Excel: Producing One Million Additional College Graduates with Degrees in Science, Technology, Engineering, and Mathematics.* Available: http://www.whitehouse.gov/sites/default/files/microsites/ostp/pcast-engage-to-excel-final_feb.pdf [April 2015].

Rothwell, J. (2013). *The Hidden STEM Economy.* Metropolitan Policy Program at Brookings Institution. Available: http://www.brookings.edu/~/media/research/files/reports/2013/06/10-stem-economy-rothwell/thehiddenstemeconomy610.pdf [April 2015].

Salzman, H. (2013). What shortages? The real evidence about the STEM workforce. *Issues in Science and Technology,* 29(4). Available: http://issues.org/29-4/what-shortages-the-real-evidence-about-the-stem-workforce/ [April 2015].

Salzman, H., and Van Noy, M. (2014). *Crossing the Boundaries: STEM Students in Four-Year and Community Colleges.* Commissioned paper prepared for the Committee on Barriers and Opportunities in Completing 2- and 4-Year STEM Degrees, National Academy of Sciences, Washington, DC. Available: http://sites.nationalacademies.org/cs/groups/dbassesite/documents/webpage/dbasse_089924.pdf [April 2015].

Schreiner, L.A., Pothoven, S., Nelson, D., and McIntosh, E.J. (2009). *College Student Thriving: Predictors of Success and Retention.* Paper presented at the Association for the Study of Higher Education, Vancouver, BC.

Schreiner, L., Kitomary, A., and Seppelt, T. (2010). *Predictors of Thriving among First-Year Students of Color.* Paper presented to the National Conference of the First-Year Experience, Denver, CO.

Shapiro, D., Dundar, A., Ziskin, M., Yuan, X., and Harrell, A. (2013). *Completing College: A National View of Student Attainment Rates-Fall 2007 Cohort.* Herndon, VA: National Student Clearinghouse Research Center.

Van Noy, M., and Zeidenberg, M. (2014). *Hidden STEM Knowledge Producers: Community Colleges' Multiple Contributions to STEM Education and Workforce Development.* Commissioned paper prepared for the Committee on Barriers and Opportunities in Completing 2- and 4-Year STEM Degrees, National Academy of Sciences, Washington, DC. Available: http://sites.nationalacademies.org/cs/groups/dbassesite/documents/webpage/dbasse_088831.pdf [April 2015].

Witham, K., Malcom-Piqueux, L.E., Dowd, A.C., and Bensimon, E.M. (2015). *America's Unmet Promise: The Imperative for Equity in Higher Education.* Washington, DC: Association of American Colleges and Universities.

2

Multiple STEM Pathways

Major Messages

- Interest in science, technology, engineering, and mathematics (STEM) credentials continues to grow.
- Students are taking complex pathways to earning STEM credentials, often transferring among institutions, entering and exiting STEM pathways at different phases of their studies, and concurrently enrolling at more than one institution.
- The make-up of the student body is not the same as 25 years ago: students are more likely to be from minority groups and to be single parents.
- "On-time" completion of a credential is infrequent: only 22 percent of students aspiring to earn a STEM degree in 4 years achieve their goal.
- The completion rates for students who aspire to a STEM degree continue to be lower than those for students in many other fields, which has led to questions about the quality of the educational experiences for STEM students.

Overall undergraduate enrollment is projected to increase in the United States in the coming decade. It has been estimated that participation in postsecondary education will rise from about 17.7 million students in 2012 to 20.2 million students in 2023 (National Center for Education Statistics,

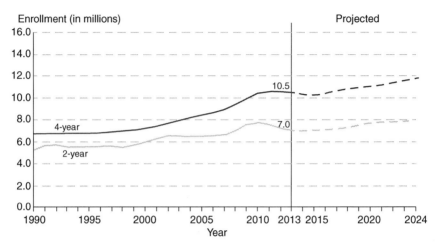

FIGURE 2-1 Actual and projected undergraduate enrollment in degree-granting institutions by level of institution: 1990–2024.
SOURCE: National Center for Education Statistics (2014, Table 303.70).

2014). This projected increase follows substantial growth in the past two decades, from about 13.0 million in 2001: see Figure 2-1.

In addition to the overall growth, enrollment has shifted across post-secondary sectors in the past two decades. Between 1990 and 2000, the growth rate for 2-year institutions exceeded those for 4-year institutions (National Center for Education Statistics, 2014). The reverse is now true: growth in 4-year enrollments now outpaces 2-year college enrollments.[1] The net result of these shifts, as shown in Table 2-1, has been relative stability in the share of students enrolled in 2-year institutions. The private for-profit sector (both 2-year and 4-year) grew rapidly between 1990 and 2013, especially between 2000–2010 when enrollments quadrupled. However, this growth is derived from a very small base of less than 2 percent of non-profit enrollments in 1990 (National Center for Education Statistics, 2014).

These trends in enrollment have occurred at the same time as changes in how students navigate the undergraduate education system. As noted in Chapter 1, the path of graduating from high school and then enrolling in a baccalaureate program and earning a bachelor's degree in 4 years is no longer the norm. An increasing number of students are earning credits from multiple institutions, are transferring between institutions (from 2-year to 4-year institutions, from 4-year to 2-year institutions, between 2-year insti-

---

[1]These trends typically correlate with economic cycles. When the economy is in decline, 2-year enrollments increase faster than 4-year enrollments; when the economy is recovering, the reverse happens.

TABLE 2-1 25-Year Changes in the Undergraduate Student Population at 2-Year and 4-Year Institutions (in percentage)

Student Characteristics	1987	2012
Aged 25 and Older	37	40
Enrolled in 2-Year Institutions	43	40
Enrolled Part Time	42	50
Minority	20	42
Employed Part Time	[a]	40
Employed Full Time	26	27
Parents	20	26
Single Parent	7	15
Women	54	57

[a] Part-time employment data were not available in 1987.
SOURCES: Data from *Digest of Education Statistics* (National Center for Education Statistics, 1990) and U.S. Department of Education, Integrated Postsecondary Education Data System (National Center for Education Statistics, 2012).

tutions, and between 4-year institutions), or are enrolled in more than one institution at the same time. In fact, in 2012, 45 percent of all bachelor's degrees were awarded to students who earned credits from a community college (National Student Clearinghouse, 2012).

TODAY'S STUDENTS

Analyses of National Student Clearinghouse data on all first-time, full-time, and part-time students who started in any type of institution in 2006 (nearly 2.8 million students) show that over approximately 5 years, one-third of the students transferred to a different institution between their initial enrollment and degree completion (Hossler et al., 2012). Most transfers took place in the second year, but there were significant numbers for all 5 years: 15 percent, first year; 37 percent, second year; 26 percent, third year; 22 percent, fourth year; and 25 percent, fifth year. The total is more than 100 percent because 25 percent of students transferred at least twice. A total of 43 percent of students who transferred from all types of institutions went to a public 2-year college, making this the most popular destination (Hossler et al., 2012). Community colleges are popular destinations for transferring students due to a number of factors, including lower cost, increased accessibility (The College Board, 2014), and proximity to students' homes, relative to 4-year institutions.

Across all fields of study, it is uncommon for students to graduate on time (e.g., completing a 2-year degree in 2 years or a 4-year degree in 4 years). The on-time completion rate for 1- and 2-year certificates is just

16 percent; for 2-year associate's degrees, it is just 5 percent;[2] and for 4-year bachelor's degrees, it is less than 35 percent (Complete College America, 2014). In addition, not all students enroll in college in the academic year after graduating from high school (Kena et al., 2014). Many students take one or more semesters off between high school and college, and some only enroll years later. Together, the array of entrance and exit points and multiple institutional enrollment patterns create a complex set of student pathways for obtaining an undergraduate credential.

Along with changes in how students navigate their way to credentials in STEM and other fields, the demographic profile of the students who are attending undergraduate institutions is also changing. Today's undergraduate college population looks somewhat different from the college population of 25 years ago (Table 2-1). For students from low-income families, there has been a nearly 18-percent increase in enrollment since 1990 (National Center for Education Statistics, 2014), and women are a slightly larger majority, about 57 percent today compared with 54 percent 25 years ago. The student population is also now more racially and ethnically diverse (National Center for Education Statistics, 2014). Increasing numbers of black and Hispanic students are attending college: as a consequence, non-Hispanic white students now account for a smaller fraction of all college students. In 1990, 77 percent of college students were non-Hispanic white; in 2012, the number was 57 percent. Between 1990 and 2012, the percentage of college students who were black rose from 12 to 15 percent, and the percentage of students who were Hispanic rose from 6 to 16 percent. During the same time period, the percentage of students who were American Indian/Alaska Native remained relatively stable (0.8% and 0.9%).

The student population is slightly older than in the past. In 2012, about 60 percent of undergraduate students were under age 25, compared with 63 percent in 1987. Today's diverse populations of undergraduate enrollees are distributed very differently across types of institutions by age and by race (National Center for Education Statistics, 2014). For example, in fall 2011, 14 percent of full-time students enrolled at 4-year public institutions were over age 25, compared with 29 percent of full-time students enrolled at 2-year public institutions. A greater proportion of part-time students at both 4-year public institution (50%) and 2-year public instructions (48%) in fall 2011 were over 25.

Changes in the student population are linked to precollege factors. The percentage of students who completed high school in 2012 differs by socioeconomic status and by race and ethnicity. The gap in access to college is also apparent in the difference in college-going rates after high

---

[2] Students who transfer to 4-year institutions without earning an associate's degree are counted against the on-time completion rate.

school graduation among students from different backgrounds. Students from families with a high income are more likely to enroll in postsecondary institutions the year after completing high school (81%) than students from middle- (65%) or low- income (52%) families (Kena et al., 2014). The higher enrollment rates of white students compared to black students first measured in 1990 no longer existed in 2012. In 2012, only Asian students enrolled in a postsecondary institution the year after completing high school (84%) at a higher rate than other students: white (67%), Hispanic (69%), and black (62%) students.[3] While the gap in enrollment in college after completing high school between racial minorities and whites has been closed, lower percentages of black (68%), American Indian/Alaska Native (68%), and Hispanic (76%) students graduate from high school compared to white students (85%) (Kena et al., 2014). In addition, students from racial minority groups continue to be concentrated in community colleges, less selective 4-year institutions, and for-profit institutions. There is some research on the precollege factors that influence student aspirations to earn STEM degrees. See Box 2-1 for an overview of factors related to engineering.

The rest of this chapter explores how these trends are reflected in the composition of the pool of students pursuing undergraduate STEM credentials and the pathways they take through the undergraduate education system. We look at the 4-year pathways, the 2-year pathways, and the for-profit sector. We discuss data regarding who completes STEM degrees and who does not. Throughout, we consider the similarities and differences among STEM aspirants and the overall undergraduate student population. Limitations in the nationally representative data sources on STEM education restricted our exploration of the array of pathways to complete a STEM credential: see Box 2-2. We close with conclusions regarding these STEM pathways.

THE 4-YEAR COLLEGE PATHWAY TO A STEM DEGREE

In the last decade, the United States has seen roughly a 10 percentage point increase in the numbers of first-time, full-time students who enter 4-year institutions with the intention of pursuing a major in a STEM discipline (Eagan et al., 2013; Hurtado et al., 2012; National Science Foundation, 2014). Although interest in pursuing STEM majors continues to increase, overall STEM completion rates have remained stagnant, and disparities among underrepresented groups persist (Eagan et al., 2014). Two previous consensus reports and a recent workshop captured this scenario and have already made the case for improvements in undergraduate STEM education, especially for students from groups typically underrepresented

---

[3] Data on American Indian/Alaska Native students were not available.

BOX 2-1
Precollege Factors that Influence Student Pathways to 2-Year and 4-Year Engineering Degrees

Although this report focuses on undergraduate education at 2-year or 4-year institutions and does not attempt to address precollege preparation, young people face many challenges bridging precollege course work and experiences and their initial foray into studying a STEM discipline. Engineering has some unique barriers to attracting and retaining undergraduate students that merit mention.

Many students select engineering as a major without actually knowing what engineering is. This lack of understanding is not specific to students: *Changing the Conversation* (National Academy of Engineering, 2008) documented that engineering as a field is not well understood by the general public. Engineering is starting to be incorporated into statewide K-12 standards and is also being introduced into some K-12 curricula and extracurricular programs. However, the implementations vary widely. In some states, engineering is embedded in science standards; in others, it is part of new standards that cover engineering and technology. In some schools, it is taught by science teachers; in others, by technical education teachers. In most schools, it is taught by teachers who have had no formal training in engineering.

Against this backdrop, students sometimes face a mismatch between their expectations and what they find when they enter an engineering program. Confronted with difficult "gatekeeper" courses in the first year of college, they often lack the bigger picture that might encourage them to continue. The other significant challenge is that many engineering programs require high school students to apply directly into a specific engineering discipline (e.g., mechanical engineering), which requires a level of knowledge and exposure that most U.S. high school students probably do not have.

---

among STEM degree earners: see National Academy of Sciences (2007); National Research Council (2011); and National Academy of Engineering and American Society for Engineering Education (2014).

Although students who begin college as traditional first-time, full-time students may have higher probabilities of attaining STEM career goals than non-first-time (e.g., transfer or returning students) and part-time students, "the "traditional" pathway of entering college as a STEM major and completing that degree program in 4 years "is becoming anything but typical or commonplace" (Eagan et al., 2014, p. 2). Many first-time students who begin at 4-year colleges and universities switch into and out of STEM majors, concurrently enroll at more than one campus, take semesters or full years off (often referred to as stopping out), and even drop out of college. These patterns differ across students' background characteristics, initial intended majors, type of institution, and where students initially enroll (Eagan et al., 2014).

BOX 2-2
Data Limitations

Three federal major statistical sources provide nationally representative information on undergraduate science, technology, engineering, and mathematics (STEM) education: the Integrated Postsecondary Education Data System of the National Center for Education Statistics at the U.S. Department of Education; the National Center for Science and Engineering Statistics of the National Science Foundation; and the American Community Survey on Educational Attainment from the U.S. Census Bureau. These statistical sources provide a wealth of critical information about undergraduate education at 2-year and 4-year institutions in general and about undergraduate STEM education in particular. Yet these sources only collect a limited amount of data related to the committee's tasks. Some of the limitations of these sources were overcome by using nonfederal data sources, such as the Cooperative Intuitional Research Program (CIRP) Freshman Survey and the National Student Clearinghouse.

Overall, however, our analysis was constrained by several factors:

- Representative data only exist on full-time, first-time students.
- Information on intended major when students are first enrolled is only available for students at 4-year institutions.
- Data on the quality of students' educational experiences are very limited.
- Data on who teaches college courses—that is, their training or qualifications—are no longer collected.
- Degree completion data only span 6 years.
- Data are not available on subgroups among Hispanics and Asian Americans.
- The sample sizes are sometimes too small for meaningful analysis for groups such as Native Americans, first-generation students, veterans, and students with disabilities.

Trends in Student Aspirations

Drawing from nationally weighted data collected from the Cooperative Institutional Research Program's (CIRP) annual Freshman Survey for 2004 (Sax et al., 2004) and matched with data from the 2010 National Student Clearinghouse (NSC), Eagan and colleagues (2014) provide trend analyses on aspiring first-time freshmen and longitudinal analyses that focus on completion rates based on the characteristics of students who intend to pursue STEM and students who were non-STEM majors at college entry. It is important to note that Eagan and colleagues include the natural sciences, technology, engineering, and mathematics as the default components of STEM; when social and behavioral sciences are included in their analyses, it is specifically noted.

The Freshman Survey covers hundreds of thousands of first-time, full-time entering freshmen at 4-year colleges and universities nationwide. The National Science Foundation (NSF) relies on these data for the National Science Board's biennial *Science and Engineering Indicators* report. The data are weighted within institution and within institutional type by gender, and the weighted data represent characteristics of the national population of first-time, full-time freshmen in nonprofit 4-year colleges and universities in the United States.

To examine persistence and completion rates of students, Eagan and colleagues (2014) matched data from the 2004 Freshman Survey with enrollment and completion data from NSC. The timeframe for the NSC data ranged from August 2004 through June 2010, which allowed for analyses regarding 4-, 5-, and 6-year degree completion for students who entered a 4-year college or university as a first-time, full-time freshman. The combined dataset also has been weighted by gender within institution and within institutional type to make this sample of first-time, full-time freshman representative of the national population of first-time, full-time students who entered college in fall 2004.

The Freshman Survey includes more than 250 variables representing student characteristics, precollege experiences, and educational and career goals. To identify the characteristics of students who intend to pursue STEM majors when they enter college, Eagan and colleagues primarily relied on student demographic characteristics, intended major, and precollege academic preparation. Tracking STEM aspirants is essential, as most studies focus on STEM students after they have declared a major and therefore underestimate the loss of STEM student talent in the first 2 years of college.[4] There is also evidence that choosing a STEM major is directly influenced by intent to major in a STEM field (Wang, 2013).

Figure 2-2 shows a slight increase from 2001 to 2011 in the proportion of all entering full-time first-year students who indicate at college entry that they have an interest in majoring in STEM. With the exception of mathematics, all STEM fields show increased student interest and have recovered in the last decades from an all-time low in the late 1980s. Comparing student intentions by race and ethnicity, the initial gap between underrepresented minority students and white and Asian students evident in 1971 has largely been closed, and only in the last few years is there evidence of slight differences, with 38 percent of white and Asian students aspiring to STEM majors, compared with 35 percent of underrepresented

---

[4]All students who aspire to a STEM degree do not start college in a STEM major. Many enter college without declaring a major or in a non-STEM major. For example, many 2-year colleges do not require students to declare a major, and students do not need to receive an associate's degree prior to transferring to a STEM major at a 4-year institution.

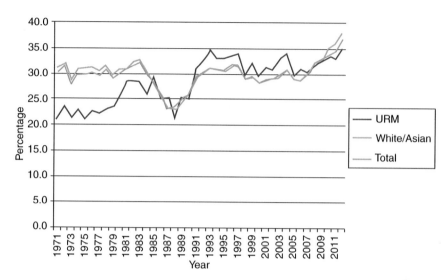

FIGURE 2-2 Percentage of first-time full-time students intending to major in STEM, 1971–2012.
NOTE: URM = underrepresented minority.
SOURCE: Eagan et al. (2014, Figure 2).

minority students. Asian American students are still slightly more likely to aspire to a STEM degree than all other groups. Hispanic students' interest has increased along with their growth in the college population.

Women's interest in STEM majors has increased substantially, along with their representation in the college population. One notable trend, illustrated in Figure 2-3, is that the gender gap has been reversed among STEM aspirants. In 1971, 62 percent of men and 38 percent of women aspired to a STEM degree; in 2012 the percentages were 48 percent and 52 percent respectively. When social sciences are included in the analysis of STEM aspirants, more than half (52%) of all first-time, full-time students indicated an interest in a STEM major. In addition, it is important to note that female aspirations to earn a STEM degree differ by discipline. Females are a big majority in social sciences (70%) and a majority in biological sciences (62%), while they are distinct minorities in engineering (21%) and in math and computer science (25%).[5]

---

[5] Women account for less than 20 percent of bachelor's degrees in computer science and more than 40 percent of bachelor's degrees in mathematics and statistics. See http://www.nsf. gov/statistics/2015/nsf15311/digest/nsf15311-digest.pdf [July 2015].

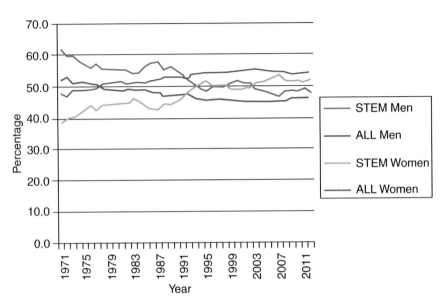

FIGURE 2-3 Percentage of first-time full-time undergraduates intending to major in STEM by gender, 1971–2012.
SOURCE: Eagan et al. (2014, Figure 4).

Student Characteristics

Students who intend to major in STEM areas differ from all students in their level of precollege preparation: STEM-interested students begin college better prepared academically, more likely to have a higher than average grade point average (GPA), and more likely to have completed higher-level courses in mathematics (including calculus and advanced placement [AP] calculus) (Eagan et al., 2014). Not surprisingly, aspiring engineers are more likely to enter college with higher levels of mathematics; those who have aspirations in the biological sciences have more years of biology; and those who have aspirations in the physical sciences have more years of high school physical science coursework (Eagan et al., 2014). Demographic differences in intended majors occur across fields: women are more likely to pursue biological sciences, health professions, and social sciences and men are more likely to intend majors in engineering, mathematics, and computer science, as well as the physical sciences. The social sciences have the greatest percentage of aspirants from historically underrepresented groups: see Table 2-2. Social science aspirants are more likely to come from low-income backgrounds (38%) than physical science aspirants (26%). More than one-third of aspirants to health professions majors come from the lowest income category.

TABLE 2-2 Student Characteristics and Precollege Preparation across STEM Disciplines and Social Sciences (in percentage)

Student Characteristics	Biological Sciences (15,338)	Engineering (15,727)	Health Professions (17,444)	Math/ Computer Science (3,850)	Physical Science (4,140)	Social Science (20,763)
Gender						
Men	40	79	25	75	57	30
Women	61	21	75	25	43	70
Race						
American Indian	<1	<1	<1	<1	<1	<1
Asian	14	13	9	16	10	7
Black	8	6	10	8	5	10
Latino	9	9	9	8	6	14
White	54	59	59	53	65	53
Other	15	13	13	15	14	15
Income						
Below $50K	30	25	34	32	26	38
$50K–$100K	30	32	34	31	34	29
Above $100K	40	43	32	37	40	33
Mother's Education						
No college	26	23	32	27	22	31
Some college	16	15	18	16	16	17
College degree or higher	59	62	51	58	62	52
Precollege Preparation						
HS GPA: A-or higher	62	62	50	55	64	45
Years of HS math: 4 or more	92	94	87	92	92	84
Years of HS physical science: 3 or more	29	39	27	33	50	28
Years of HS biological science: 3 or more	29	12	23	13	16	18
Completed calculus	39	51	25	45	45	24
Completed AP calculus	42	60	21	51	50	22

NOTES: AP = advanced placement; GPA = grade point average; HS = high school.
SOURCE: Eagan et al. (2014, Table 2).

Completion Rates

The majority of students who enter a 4-year institution intending to major in the natural sciences, technology, engineering, and mathematics do not earn a degree in these fields, and most of the students who switch majors do so after an introductory course in mathematics, science, or engineering (President's Council of Advisors on Science and Technology, 2012). There is some evidence suggesting that many students who perform well in introductory classes and are capable of earning a STEM degree still switch majors (Seymour and Hewitt, 1997; Brainard and Carlin, 1998). Students who are interviewed about why they switched majors often cite uninspiring and ineffective classroom environment and teaching practices as the reason (Seymour and Hewitt, 1997). The population of those who complete STEM degrees is argued to be the result of the cumulative effects of students' individual decision making in response to factors in their institutions (e.g., quality of teaching, availability of support structures, discovery of attractive alternative majors) and external factors (e.g., early educational preparation, financial concerns, and larger social issues that affect specific groups).

STEM degree completion varies across fields, by students' race, ethnicity, and gender, and by institutional type (Eagan et al., 2014). Figure 2-4 shows the probability of completing the originally intended major, switch-

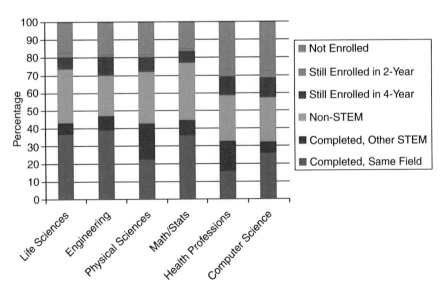

FIGURE 2-4 Six-year enrollment and completion status of first-time full-time STEM aspirants at 4-year institutions who began in 2004, by initial STEM field.
SOURCE: Eagan et al. (2014, Figure 6).

ing to another STEM field, switching to a non-STEM field, still being enrolled after 6 years, and no longer being enrolled in college, all by students'
initial field of study. Engineering and life science programs appear to do
a better job of retaining students: 39 percent of engineering, 37 percent
of life science, and 36 percent mathematics aspirants completed a degree
in that field in 6 years, and another 8 percent, 6 percent, and 8 percent
respectively switched to a different STEM field. The reason for this difference for engineering and life science aspirants is unclear. For engineering
aspirants this trend could be due to the higher academic characteristics of
aspiring engineers (evidenced in Table 2-2, above), the timing of entry into
an engineering major (sometimes occurring in the third year of college), or
other factors.

In contrast with engineering and life sciences, less than 25 percent of
students who began college intending to major in the physical sciences
completed a degree in 6 years, 20 percent shifted to a different area of
STEM, and nearly 30 percent switched to a non-STEM major. Mathematics
and statistics lost the largest percentage of their aspiring majors to non-
STEM fields (32%), but their aspirants were more likely to complete a
bachelor's degree in any field (67%) and less likely to have dropped out of
higher education (15%) (Eagan et al., 2014).

Not all STEM degree earners state an interest in a STEM degree when
entering college. Among the 34,616 students who earned a STEM degree
in the dataset analyzed by Eagan and colleagues, 18 percent originally
intended to pursue a non-STEM major. About 30 percent came from
the group of students who originally indicated they were "undecided/
undeclared" at college entry. Fields from which the largest numbers of
students who switched into a STEM major were drawn were business
(16%) and education (14%).

Completion rates vary considerably by race and ethnicity, gender, and
STEM fields. Although historically underrepresented racial minority students now aspire toward STEM degrees at the same rates as white and Asian
American students, disparities in STEM completion by race and ethnicity
persist: see Figure 2-5. First, overall, students are taking more time for the
degree—typically 5 years: only 22 percent of initial STEM aspirants completed a STEM degree in 4 years. Within 6 years of entering college in 2004,
just over 40 percent of all first-time, full-time STEM aspirants completed a
STEM degree. Within this cohort Asian American students outpaced their
peers in STEM at the 4-, 5-, and 6-year completion rates, with a total of
52 percent completing a STEM degree in 6 years. White students lagged
their Asian American counterparts, with 43 percent completing a STEM
bachelor's degree in 6 years. Historically underrepresented minorities lagged
further, with only 29 percent of Hispanic aspirants, 25 percent of American
Indian aspirants, and 22 percent of black aspirants earning a STEM degree

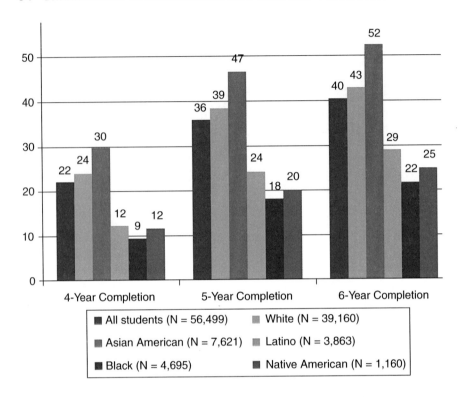

FIGURE 2-5 Cumulative percentage of 2004 STEM aspirants who completed STEM degrees in 4, 5, and 6 years.
SOURCE: Eagan et al. (2014, Figure 7).

in 6 years. By comparison, the 6-year completion rates are higher across all majors for Hispanics, American Indians, and blacks: 29 percent, 25 percent, and 22 percent respectively (Eagan et al., 2014).

The completion rates by gender and field for STEM aspirants are shown in Table 2-3. Interestingly, the 4-year completion rate was nearly the same for women and men (23% for women and 21% for men), but the rate was lower for women after 6 years (38% for women and 43% for men). The 6-year completion rates vary across fields, with women aspirants more likely than men to complete engineering degrees (43% of women and 40% of men); male aspirants more likely than women to complete bachelor's degrees in the physical sciences (28% of women and 33% of men); and male and female aspirants in the biomedical sciences about equally likely to complete the bachelor's degree (34% of women and 34% of men).

Degree attainment rates among initial STEM aspirants also vary by

TABLE 2-3 Cumulative Percentage of STEM Aspirants at 4-Year Institution Who Completed a STEM Degree in 4, 5, or 6 Years after Entering College in 2004 (N = 56,499)

Discipline and Completion Time	Men	Women
4-Year STEM	21	23
5-Year STEM	37	34
6-Year STEM	43	38
4-Year Engineering	15	20
5-Year Engineering	34	40
6-Year Engineering	40	43
4-Year Biomedical Sciences	23	22
5-Year Biomedical Sciences	32	32
6-Year Biomedical Sciences	34	34
4-Year Physical Sciences	23	23
5-Year Physical Sciences	31	27
6-Year Physical Sciences	33	28

SOURCE: Eagan et al. (2014, Table 4).

the type of institution.[6] Doctoral and research universities outperformed liberal arts and master's comprehensive institutions for STEM completion rates among STEM aspirants in engineering and biomedical sciences, but liberal arts colleges outperformed the other institutions when considering completion in the physical sciences (Eagan et al., 2014). Although private institutions had a completion advantage over public institutions, a previous study indicates that the differences in completion rates become nonsignificant after accounting for differences in the types of students enrolled at public and private institutions and for resource disparities across these institutions (Hurtado et al., 2012).

STEM completion rates differ across predominantly white institutions, historically black colleges and universities, Hispanic-serving institutions, and emerging Hispanic-serving institutions:[7] see Table 2-4. The emerging Hispanic-serving institutions showed the highest completion rates for STEM bachelor's degree aspirants at 4 years (27%), 5 years (44%), and 6 years

---

[6]The Carnegie Classification System for Institutions of Higher Education was used in the following analyses. For an overview of the Carnegie Classification System of Institutions of Higher Education, see http://carnegieclassifications.iu.edu/ [October 2015].

[7]Emerging Hispanic-serving institutions are those enrolling 15-24 percent Hispanics, just below the 25 percent cutoff for Department of Education designation as a Hispanic-serving institution.

TABLE 2-4 Cumulative Percentage of STEM Completion by Minority-Serving Institution Status (N = 56,499)

Student Population Served	Cumulative Completion Rate		
	4 Years	5 Years	6 Years
Predominantly White Institutions	23.7	38.0	42.6
Historically Black Colleges and Universities	8.0	15.6	19.3
Hispanic-Serving Institutions	10.0	22.2	28.6
Emerging Hispanic-Serving Institutions	26.7	44.1	47.5

SOURCE: Eagan et al. (2014, Table 5).

(48%). Completion rates in STEM majors were lower at Hispanic-serving institutions and historically black colleges and universities. Eagan and colleagues (2014) found that these institutions typically enroll larger numbers of students from low-income, first-generation, and underrepresented groups who have lower completion rates at many colleges and often do not have the same level of resources as students at selective predominantly white institutions. When Eagan and colleagues (2014) controlled for student and institutional factors, they found that the difference in completion rates between minority-serving institutions and predominantly white institutions became nonsignificant. In addition, these multivariate analyses demonstrated that black STEM aspirants are more likely to graduate with a STEM degree if they attended a historically black college or university than if they had been enrolled at a predominantly white university.

Student Mobility

Enrollment mobility is often unaccounted for in discussions of STEM students. Mobility is highest among traditional-age college students who begin at 4-year institutions. Eagan and colleagues' (2014) analysis of aspiring STEM majors' trajectories found that, over six years, approximately 15 percent of these students reverse transferred from 4-year to 2-year institutions; 13 percent transferred laterally from one 4-year institution to another; and approximately 9 percent were concurrently enrolled in more than one institution (or campus) (Salzman and Van Noy, 2014). Data on first-time college students (which is not limited to full-time freshmen) from the Beginning Postsecondary Student (BPS) Survey of the National Center for Education Statistics, indicate that 42 percent of 4-year STEM degree holders reported they had reverse transferred, laterally transferred, or were concurrently enrolled (Salzman and Van Noy, 2014). A separate longitudinal study found that between 2001 and 2007, about one-half of all STEM bachelor's degree recipients had attended a community college at some point in their college career (Mooney and Foley, 2011).

Attending multiple institutions is associated with increased time to degree and lower STEM degree completion rates (Salzman and Van Noy, 2014). The relationship between STEM student mobility and completion rates is shown in Table 2-5: low levels of completion are associated with reverse transfers and slower progression with lateral transfers. Concurrent enrollment was not as strongly related to students' completion as transfer. Van Noy and Zeidenberg (2014) also note a negative relationship between student mobility across 2-year and 4-year colleges and completion rates (see next section for discussion of 2-year colleges). The mobility described here hints at some of the many ways that students navigate the higher education system. It also shows the difficulties of developing metrics to track students along these multiple pathways or to assess institutions' contribution to or detraction from these students' success.

The complex picture that emerges from the analyses of 4-year college students is characterized by the following:

- strong intention to major in STEM by students from all population groups;
- different distributions across STEM fields by different demographic groups;
- losses of intended majors from STEM and recruitment to STEM of non-STEM-intending students; and
- use of multiple institutions and pathways during matriculation.

Box 2-3 provides an example of the complexity of the pathways for STEM degrees for engineering.

TABLE 2-5 Cumulative Percentage of STEM Completion by Mobility Status

Kind of Mobility	Cumulative Completion Rate		
	4 Years	5 Years	6 Years
Reverse Transfer	1	3	6
Lateral Transfer	6	17	24
Concurrent Enrollment	17	31	36
All Students	22	36	41

NOTE: The completion rates are cumulative.
SOURCE: Eagan et al. (2014, Table 8).

BOX 2-3
Engineering Pathways

Education for the profession of engineering is a complex endeavor, with students learning mathematics, science, design, and concepts from the social sciences and humanities, as well as a range of professional skills. Engineering programs vary in terms of their emphasis on engineering science, experiential education, design, and involvement in research. While some programs are innovative and provide cutting-edge education to their students, others rely on approaches to teaching that have been used for decades (Sheppard et al., 2008). There is also an acknowledged gap between the kinds of pedagogies and learning environments that engineering educators say they value and those that are currently being practiced (American Society for Engineering Education, 2012). Here we provide a description of some of the complex issues that arise in 4-year engineering pathways and the overarching issues they raise.

Progress toward degrees in engineering programs is affected by both admissions policies and curricula. At some universities, students apply to engineering at the end of their sophomore year. However, it is more common to admit students to engineering at the time they are admitted to the university, and even more common to admit them to a specific major within engineering. Therefore, even at universities that have fairly flexible entry points for other majors, engineering students may be "behind" if they do not start the engineering curriculum in their first term of college or if they struggle in their first-year courses. However, this appears to manifest itself in time to degree and migration rates for students into engineering, rather than in retention rates. As shown in Figure 2-4, the persistence of students in engineering—measured by the percentage of students who start in engineering and are still enrolled in engineering in their eighth semester—is higher than for other STEM and non-STEM disciplines (Eagan et al., 2014; Ohland et al., 2008). In contrast, the rate of migration into engineering programs is less than for other disciplines: only 7 percent of eighth-semester engineering students migrated into engineering, compared with 30–60 percent in all other majors (Ohland et al., 2008). The net result is therefore a decline in the number of engineering students between admission and degree completion. This lack of migration into engineering highlights the important role that initial choice of major plays for undergraduate engineering students and suggests that creating new entry points might increase overall completion rates.

The structure of engineering curricula may play a role in retention in the discipline, as well as the experience for students. Research shows that retention in engineering from the first year to the second year increases when a lecture-based "Introduction to Engineering" course is replaced with a hands-on course in which students do team-based projects in the context of "real" engineering problems, which can range from toy design to industry-based projects to community solutions to grand challenges (Freeman et al., 2014; Hoit and Ohland, 1998; Lichtenstein et al., 2014).

The introductory engineering course grows in importance because it is often the only engineering course that students take early in their college careers. A majority of the first-year required courses in an engineering plan of study are in mathematics and science; few are taught within a college of engineering. In some programs this absence of engineering-specific courses extends through the sophomore year. Students may therefore spend 1–2 years majoring in engineering without getting a sense of what engineering courses will be like. Compounded with the gatekeeper role that first-year mathematics courses often play, engineering students can find themselves struggling, yet without a compelling picture of why they are studying engineering.

There has been a consistent push to increase the amount of knowledge and number of skills expected from an undergraduate engineering education (e.g., National Academy of Engineering, 2004). This push has led to requirements for a large number of courses, with credit hours often higher than required in other disciplines and curricula that are sparse in free electives. In turn, this has led to growing conversations around the question of how students will learn all that is needed for 21st-century careers. Would repositioning engineering as a 5-year degree (see National Academy of Engineering, 2005) relieve some of the pressure on students or would it deter students from selecting engineering? Would redefining the bachelor's degree in engineering to be a "pre-professional" degree, similar to pre-med or pre-law (see, e.g., the Raise the Bar initiative; Russell and Lenox, 2013), increase interest or discourage students who don't want to commit to graduate study?

One of the ways in which the field has responded to the increasing expectations for engineering graduates has been to increase the curricular emphasis on design, problem-based learning, and experiential education. These are proving to be effective means of teaching both technical content and broader skills (Dym et al., 2005; Freeman et al., 2014; Lichtenstein et al., 2014; Sheppard et al., 2008; Smith et al., 2005). The outcome is programs that connect education with "the real world" through such activities as internships, service learning, research experiences, design competitions, entrepreneurship experiences, and study abroad.

Student pathways can vary substantially across the colleges and universities throughout the United States. The Academic Pathways Study, an in-depth exploration of engineering student experiences, found that these variations stem from many factors that provide both opportunities and challenges for students (Atman et al., 2010) The Academic Pathways study and other research (e.g., Lichtenstein et al., 2009; Sheppard et al., 2014) indicate that, on the positive side, graduating students are on the path to establishing their identities as engineers. They have obtained an important set of knowledge and skills, including the ability to apply concepts from mathematics and science to solve problems. They have learned to take on substantial engineering design challenges. And they have gained confidence in the kinds of professional and practical skills they will need on the job.

However, challenges also exist. Some students report heavy workloads in a competitive environment, which can be a substantial source of stress for them (Atman et al., 2010). Upper-level courses often include a focus on group projects and teamwork. Although this is becoming more common in first-year engineering courses as well, this can be a rough transition for students at institutions in

continued

BOX 2-3 Continued

which early courses still emphasize more traditional individual problem solving. A survey of engineering faculty (American Society for Engineering Education, 2012) indicates that, although there are many innovations in engineering education, a majority of faculty are far more comfortable with long-standing learning environments—such as labs, industry internships, research experiences, and competitions—than they are with newer approaches, such as service learning, entrepreneurship, and international experiences. It is therefore perhaps not surprising that many students report feeling ill prepared to incorporate broad contextual issues, including global and societal issues, in engineering problem solving. There is little room in crowded curricula for students to take advantage of study abroad programs or to study a second language. Moreover, students from underrepresented groups, including women, report different experiences than majority students, even though they are in the same classes (Atman et al., 2010). These differences can lead to lower confidence and an increased sense of work overload compared with males and majority students (Atman et al., 2010; Ohland et al., 2008).

In response to these challenges, *Changing the Conversation* (National Academy of Engineering, 2008) recommended that, as a field, engineering should talk less about the skills needed to be an engineer and more about the impact that engineering has on the world. This recommendation continues to be relevant in both recruiting and retaining students in engineering (National Academy of Engineering, 2008, p. 5):

> From research to real-world applications, engineers constantly discover how to improve our lives by creating bold new solutions that connect science to life in unexpected, forward-thinking ways. . . . We are counting on engineers and their imaginations to help us meet the needs of the 21st century.

THE COMMUNITY COLLEGE PATHWAY
TO A STEM CREDENTIAL

Community colleges are accessible and affordable, serve a diverse population, and offer a great variety of degree programs and pathways in STEM for high-skill as well as middle-skill jobs. Yet, the research base on community colleges is more limited than that for 4-year institutions. The data we reviewed indicated that community colleges play a substantial role in addressing workforce needs and in further developing the talent pool of students who may later obtain advanced STEM degrees.

Van Noy and Zeidenberg (2014) drew on the NCES BPS 2004 and 2009 surveys—which included a nationally representative cohort of students who enrolled in postsecondary education for the first time in 2003–2004 in credit-bearing programs—to analyze the pathways of community college students aspiring to earn a STEM credential. Focusing their analysis on the characteristics of community college students who enroll in STEM programs, the authors included both general STEM fields and specialized career-focused STEM programs. They included biology, mathematics, engineering, physical sciences, computer and information systems, engineering, and programs for engineering technicians, technicians, agriculture, and science technologies.[8] The major focus of their analyses was on natural sciences, engineering, technology and technician programs, and mathematics. The authors identified whether a student was in a STEM program using BPS data on student majors collected through student interviews and student transcripts.

Degree Programs

Community colleges play a significant role in STEM education. As noted earlier in this chapter, 2-year institutions played an important role in 2012, when 2-year students accounted for 40 percent of all undergraduates across all fields of study (Table 2-1). Van Noy and Zeidenberg's (2014) analysis of data on community college entrants in 2003–2004 found that about half were enrolled at some time in a STEM field over the following 6 years.

Community colleges offer two major categories of STEM programs: science and engineering programs (and a small number of mathematics programs) and technician programs. The first set of programs are transfer programs, to prepare students to pursue study that usually requires a bachelor's degree or higher; the second set are occupational programs with the goal of a credential, usually a certificate or associate's degree. These programs provide an "on ramp" to further science and engineering study, 2 years of preparation and access to an associate's degree in arts or sciences leading to transfer to a 4-year institution's program of study. Although technician programs can also lead to a degree (e.g., associate in applied science) and to transfer, their primary goal is to develop the knowledge and skills required to directly enter the workforce.

Table 2-6 breaks down enrollment in community colleges by these various programs, among students who ever enrolled in the 6 years after entry in 2003–2004. As shown in Table 2-6, about one-half of community college students enrolled in a STEM field, including science and engineering (7%), technician programs (10%), social sciences (11%), and health professions that required extensive science and mathematics coursework (23%).

---

[8] They also looked separately at programs in the social sciences and health professions, as was done for 4-year institutions (above), since the health professions have significant science and mathematics course requirements.

TABLE 2-6 Community College Enrollments by Program, Ever Enrolled in the 6 Years after College Entry among First-time Students Who Began College in 2003–2004

	Number of Students	Percentage of Students
Science & Engineering Programs		
Total science and engineering	109,592	6.6
Biological and biomedical sciences	42,152	2.6
Engineering	34,530	2.1
Physical sciences	23,776	1.4
Mathematics and statistics	9,134	0.6
Technician Programs		
Total technician	167,829	10.2
Engineering technologies	43,631	2.6
Computer and information sciences	101,264	6.1
Science technologies/technicians	5,357	0.3
Agriculture	17,577	1.1
Closely Related Programs		
Total health professions and related programs	372,721	22.6
Total social sciences	175,397	10.6
Non-STEM		
Total non-STEM	824,390	50.0
TOTAL	1,649,929	100.0

SOURCE: Van Noy and Zeidenberg, (2014, Table 1)

Comparing enrollments by type of institution, 4-year colleges had a higher representation of students majoring in science and engineering than 2-year colleges, especially for biology and engineering (Van Noy and Zeidenberg, 2014; see Figure 2-6). Conversely, 2-year colleges outpaced 4-year colleges in enrollment of engineering technician and computer and information sciences programs, reflecting the greater emphasis on workforce preparation programs in community colleges. Students' credential goals also reflected the different program orientation at 2-year and 4-year institutions. An associate's degree or certificate was the goal of 35 percent of the technician students, compared to 15 percent of the students in science and engineering programs. Sixty percent of the technician students and 80 percent of science and engineering students reported that their ultimate goal was to obtain a bachelor's degree.

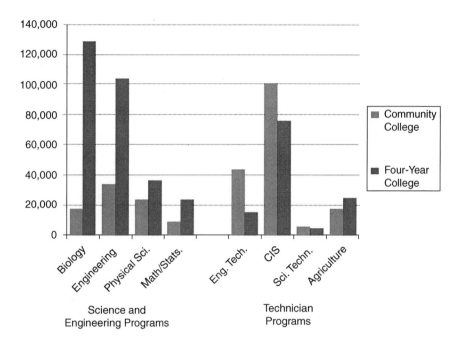

FIGURE 2-6 Number of students enrolled in STEM by program among 4-year and community college students.
SOURCE: Van Noy and Zeidenberg (2014, Figure 1).

Student Characteristics

Community college students in both science and engineering programs and technician programs shared some characteristics that distinguish them from 4-year college students: they were older and more likely to be first-generation college students; they were more likely to be working while enrolled, and when working, to work more hours than those 4-year college students who worked; and they were more likely to require developmental education (see Table 2-7). For the student populations at 2-year institutions, technician students were older than science and engineering students, included more first-generation students, and were more likely to take developmental courses than science and engineering students.

There were significant demographic differences in the students who enrolled in 2-year and 4-year institutions (Van Noy and Zeidenberg, 2014). Hispanic students were more likely to be enrolled in community colleges than in 4-year institutions in both STEM and non-STEM programs. Among 2-year STEM aspirants, Hispanic, Asian, and female students were more likely to be enrolled in science and engineering programs than in technician

TABLE 2-7 Characteristics of STEM Students at 2-Year and 4-Year Institutions (in percentage)

Student Characteristics	2-Year Students				4-Year Students	
	All STEM	Science and Engineering	Technician	Non-STEM	STEM	Non-STEM
Race/Ethnicity						
White	65	61	68	60	67	71
Black	11	8	13	15	9	10
Hispanic/Latino	14	15	12	16	9	10
Asian	6	11	4	4	9	5
All other	4	5	4	5	5	5
Female	30	40	24	62	37	62
Pell Grant Recipients	26	24	27	29	26	28
First-Generation College Student	68	62	72	73	38	46
Disabled	12	10	14	11	7	8
Age						
18–22	72	83	66	65	95	92
22–40	23	16	27	26	4	6
40+	5	1	8	8	0	2
Average Age at Enrollment	22	20	23	24	19	20
Dependent Children	17	12	19	26	2	5
Veteran	4	1	6	3	1	0
Working While Enrolled	76	78	74	78	55	62
Average Hours Worked (of those working)	30	28	30	30	19	1
Developmental Education in First Year						
Any	69	64	72	68	31	39
Math	59	56	61	59	23	31
English	14	13	15	18	6	8
Reading	15	15	16	19	4	6

SOURCE: Van Noy and Zeidenberg (2014, Table 3).

programs. Black students constituted a larger share of those enrolled in technician programs (13%) than of those enrolled in science and engineering programs (8%). Technician program enrollments were overwhelmingly white and male. Women were less likely to be enrolled in technician programs (24%) relative to the proportion enrolled in science and engineering (40%) or non-STEM programs (62%).

Community colleges are more accessible to many students because of the cost of attendance relative to that of 4-year institutions. The average price of attendance in the first year among STEM students at community college was $6,896, in comparison with $18,885 for STEM students at 4-year institutions (see Table 2-8). The expected family contributions for STEM students at 4-year institutions were higher as well: $13,987 for 4-year STEM students and $9,748 for community college STEM students.

A related difference is in loans: as shown in Table 2-8, STEM students in 4-year institutions were more likely to take out student loans while in college than students in community colleges, 62 percent and 47 percent, respectively. They also had higher student loans 6 years after their initial enrollment: the average was $21,143 for 4-year students and $15,245 for community college students.

Enrollment Patterns and Student Mobility

The enrollment patterns of STEM students at 2-year and 4-year institutions differed greatly in the BPS sample analyzed by Van Noy and Zeidenberg (2014). As shown in Table 2-9, STEM students at 2-year institutions were less likely than those in 4-year colleges to be enrolled full time (33% and 68%, respectively), and they were less likely to have had continuous enrollment with no dropouts (47% and 71%, respectively). On the other hand, students in technician programs at 2-year institutions were more likely to attend only one institution (59%) than students in science and engineering programs (33%).

Studies of all community college students, regardless of their field of study, have illustrated connections between enrollment patterns and student outcomes. These studies reveal a positive connection between continuous enrollment in community college, without multiple breaks or movement across multiple institutions, and completion of a college credential (Crosta, 2014; Goldrick-Rab, 2006). These studies also found a positive association between enrollment intensity (the amount of credit hours taken each semester) and likelihood of transfer to a 4-year institution, when transfer is the student's goal.

The frequency of student "swirling"—movement between multiple institutions prior to degree attainment—is about the same for both community college and 4-year college STEM students.

TABLE 2-8 Financial Characteristics of STEM Students at 2-Year and 4-Year Institutions

Financial Characteristics	Students at 2-Year Institutions				Students at 4-Year Institutions	
	All STEM	Science and Engineering	Technician	Non-STEM	STEM	Non-STEM
Price of Attendance in First Year	$6,896	$6,807	$7,219	$6,601	$18,885	$17,957
Expected Family Contribution in First Year	$9,748	$10,079	$9,105	$8,241	$13,987	$13,045
Percentage with Student Loans after 6 Years	47%	45%	52%	40%	62%	64%
Average Student Loan among Those with Loans after 6 Years	$15,245	$14,163	$17,007	$13,438	$21,143	$21,042

SOURCE: Van Noy and Zeidenberg (2014, Table 4).

TABLE 2-9 Enrollment Patterns of STEM Students, by Subfield, at 2-Year and 4-Year Institutions (in percentage)

Enrollment Patterns	Students at 2-Year Institutions				Students at 4-Year Institutions	
	All STEM	Science and Engineering	Technician	Non-STEM	STEM	Non-STEM
Average Enrollment Intensity						
Always full time	33	36	32	27	68	65
Always part time	13	8	15	22	1	2
Mixed part time and full time	53	55	53	51	31	33
Constancy of Attendance/Number of Stopouts						
0	47	49	46	50	71	72
1	41	43	39	35	22	21
2+	12	8	15	15	7	7
Institutional Attendance						
Attend only one institution	49	33	59	62	75	74
Traditional transfer	25	41	16	19	NA	NA
Attend multiple institutions, swirling	26	26	25	19	25	26

SOURCE: Van Noy and Zeidenberg (2014, Table 5)

About the same proportion of community college students are likely to switch into STEM fields after initial enrollment as the proportion of entrants indicating a major in STEM, as shown in Table 2-10. Possible explanations for this later entry include limited advising capacity of the institution, indecision related to the lack of exposure to options, or the regular process of career exploration (factors that are not unique to community college institutions and students). There are consequences to such delaying selection of a major, including extended time to completion and increased cost (Van Noy and Zeidenberg, 2014). It is important to note that this analysis of selection of a major does not capture the major that students aspired to earn when starting college, because this information is not captured in the BPS survey. Thus, the loss of STEM aspirants prior to declaring a major is not represented in this analysis.

Community college STEM students switch out of STEM at a higher rate than 4-year students in STEM majors (28% and 22%, respectively; for more details, see Table 2-10). They also take more developmental courses, especially in mathematics, than students at 4-year institutions (Van Noy and Zeidenberg, 2014). Students who switch fields of study move into a range of non-STEM majors, including business, health professions, and education. There may be at least two possible interpretations of these switches. Some students may discover that they do not like the STEM program or have found a program that is a better match for their interests and abilities: if so, their departure from STEM is not a negative outcome but rather part of the natural process of exploration and discovery in college. Another interpretation is that some students have negative experiences in STEM programs for which they are otherwise actually a good match: if so, it would be a major concern. Existing research points to the fact that the culture of STEM classrooms and departments are unwelcoming to many students, especially women and underrepresented minorities (Ramsey et al.,

TABLE 2-10 Major Decision Making among STEM Students (in percentage)

| Major Decisions | Community College | | | 4-Year College |
	All STEM	Science and Engineering	Technician	All STEM
Timing of Entry into STEM				
Enter STEM at initial enrollment	51	53	51	62
Switch into STEM after first year of enrollment	49	47	49	38
Switch out of STEM to a Non-STEM Major	28	27	28	22

SOURCE: Van Noy and Zeidenberg (2014, Table 6).

2013; Seymour and Hewitt, 1997). Some departmental, institutional, state, and federal policies may also serve to push students away from attaining a STEM degree. We explore the effects of these and other barriers on student completion of STEM degrees in Chapters 3 and 4.

Degree Attainment

Given students' varied intentions and credential goals, Van Noy and Zeidenberg (2014) warn against the sole focus on degree completion, cautioning that multiple measures of community college STEM outcomes are necessary. As discussed in Chapter 1, students at 2-year colleges may seek to earn a 2-year degree, transfer to a 4-year institution without earning a degree, earn a certificate, or learn job-related skills. Thus, in addition to measures of credential completion, other measures of transfer, credential attainment at other institutions, continued enrollment, and employment are needed to assess community college student outcomes (Rassen et al., 2013).

About 30 percent of STEM community college students had either earned a credential or were still enrolled in STEM, and about 33 percent had either attained a credential or were still enrolled in a non-STEM field (Van Noy and Zeidenberg, 2014; see Table 2-11). Of those who left STEM, students in technician programs had a very different trajectory from those in science and engineering programs: for example, they were more likely to have left college without completing any credential (41% in science and 27% in engineering). The lower rate of completion among students in technician programs may be due to obtaining employment prior to completing the requirements for a credential. Or it may be due to any number of negative factors, such as insufficient money to proceed. Without reliable data on why students leave college prior to completing a certificate or degree, it is not possible to gauge the success of these technician programs.

In terms of degree outcomes, about 20 percent of STEM community college students attained any STEM credential 6 years after enrollment (see Table 2-12). Sixteen percent of science and engineering students and

TABLE 2-11 Community College Student Completion and 6-Year Retention Rates (in percentage)

Outcome	All STEM	Science and Engineering	Technician
Attained Credential or Still Enrolled in STEM	30	33	30
Attained Credential or Still Enrolled in Non-STEM	33	39	29
Dropped Out without Credential	37	27	41

SOURCE: Van Noy and Zeidenberg (2014, Table 7).

TABLE 2-12 Six-Year Outcomes for Community College STEM Students (in percentage)

Outcome	All	Science and Engineering	Technician
Attained STEM Credential			
Any credential	19	21	20
Bachelor's	10	16	7
Associate's degree or certificate	9	5	13
Still Enrolled			
At any institution	16	19	14
At community college	7	6	8
At 4-year college	8	13	6
Transferred to 4-Year College in STEM Program	25	37	19

NOTE: Students may be included in more than one category; students who transferred may also be counted as attaining a STEM credential or still enrolled in a STEM program.
SOURCE: Van Noy and Zeidenberg (2014, Table 8).

7 percent of technician students completed STEM bachelor's degrees. In addition, 16 percent of all STEM students were still enrolled in STEM 6 years after initial enrollment (19% of science and engineering students and 14% of technician students).

THE FOR-PROFIT SECTOR PATHWAY TO A STEM CREDENTIAL

The for-profit sector of postsecondary education differs from the non-profit (public and private) sector in three essential ways: finance, governance, and a market-driven focus. The distinguishing feature of for-profit institutions is that they are businesses, ranging from small family-owned activities to large corporate entities, which are run to generate revenues. These institutions are accountable to investors and stockholders, as well as to state and federal governments, and they have a strong customer service orientation (Ruch, 2001).[9] The for-profit institutions have the capacity to move swiftly to meet market demand in growing STEM areas.

Degree Programs and Attainment

Many for-profit institutions offer certificates and nondegree training, and they also award accredited associate's, bachelor's, and graduate degrees. They are usually accredited by a national accreditor rather than the regional

---

[9]See Kinser (2014) for a history, scope, and diversity of the institutions.

TABLE 2-13 Completions in STEM Fields in 2012

Fields	Public	Private Nonprofit	Private For Profit
Health Professions and Related Programs	401,479	97,544	330,964
Computer and Information Sciences and Support Services	64,906	15,462	38,597
Engineering Technologies and Engineering-Related Fields	59,952	4,361	26,088
Engineering	68,353	20,049	382
Biological and Biomedical Sciences	72,452	32,122	201
Science Technologies/Technicians	3,514	188	100
Physical Sciences	23,040	9,021	27
Mathematics and Statistics	15,976	7,811	1
TOTAL	709,672	186,558	396,360

SOURCE: Kinser (2014).

accreditors that service the nonprofit institutions (Kinser, 2014).[10] Many offer credentials in STEM fields, often for middle-skills jobs (not requiring a 4-year degree) for which growth is projected and student demand is high. In 2012, for-profit institutions awarded slightly less than half the number of STEM credentials awarded by nonprofit institutions (both public and private), as shown in Table 2-13.

Across all types of postsecondary institutions, credentials in the health professions were the most frequently awarded (Kinser, 2014; Table 2-13). However, there were striking differences between for-profit and nonprofit institutions in the concentration of programs of study and the types of credentials awarded. First, more than 80 percent of the credentials awarded by for-profit institutions were in health professions and related programs, compared with just over 50 percent of the credentials awarded by public and private nonprofit institutions (Kinser, 2014). The for-profit sector also awards large numbers of engineering, technology and computer and information science credentials. Second, bachelor's degrees made up a much smaller proportion of the total STEM credentials awarded by for-profit institutions compared to nonprofit institutions (see Figure 2-7) (Kinser, 2014).

Still, the numbers of graduates and scale of the for-profit sector are significant. In 2012, for-profit institutions awarded around 35,000 bachelor's degrees, 102,000 associate's degrees, and 257,000 certificates in STEM fields (Figure 2-7). For-profit institutions offer many online degree pro-

---

[10]The data on for-profit institutions analyzed by Kinser are from NCES.

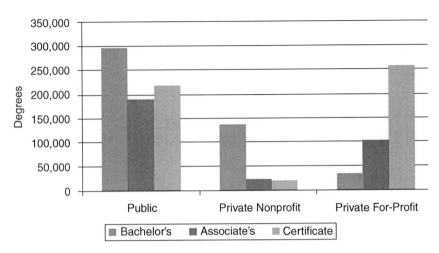

FIGURE 2-7 Number of degrees awarded in STEM fields at public, private nonprofit, and private for-profit institutions, in 2012.
SOURCE: Kinser (2014).

grams and Internet course delivery that is convenient to different groups of students, especially those who are working full time. In 2012, the University of Phoenix online campus—the largest postsecondary institution in the United States—awarded 20,798 STEM credentials, mostly associate's and bachelor's degrees in the health professions. It has also added new STEM fields of study (e.g., computer networking, security, and administration).

Student Characteristics

The for-profit institutions train a diverse population of students who take varied pathways to a STEM credential. In 2012, about half of all STEM credentials earned by black, Hispanic, and native Hawaiian and other Pacific Islanders were from for-profit institutions (Kinser, 2014; see Figure 2-8). For-profit institutions typically attract students whose goal is to "get in, get out, and get a job." Recent analyses by the U.S. Department of Education (National Center for Education Statistics, 2015) indicate that, in the fall of 2013, students enrolled in for-profit institutions (both 2-year and 4-year and both full time and part time) were older than comparable students at nonprofit 2-year and 4-year institutions. Earlier data suggest that the majority of students at for-profit institutions work 35 or more hours per week (Ruch, 2001, p. 134).

According to the Institute for College Success and Access (2014), 88 percent of students in for-profit institutions graduate with student debt (averaging $39,950), compared with 75 percent of students in private non-

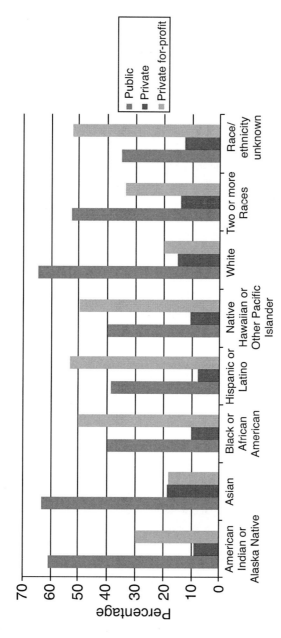

FIGURE 2-8 STEM credentials awarded from public, private nonprofit, and private for-profit, by students' race and ethnicity, in 2012. SOURCE: Kinser (2014).

profit institutions (averaging $32,300) and 66 percent of students in public institutions (averaging $22,550). Although most students at for-profit institutions are studying in programs at the sub-baccalaureate levels, these data raise an important set of questions that remain unanswered: Why do these students pursue a for-profit education in STEM fields even though typically the costs are higher for them? Is it the promise of a job and short degree program or convenience of an online education? How do these nondegree and degree holders fare in the job market? Is the curriculum too narrow to allow movement from for-profit to nonprofit degree programs? The answers to these questions could be instructive for nonprofit institutions working on diversifying the STEM fields and possibly result in articulation agreements to align for-profit with nonprofit postsecondary education curriculum and training goals.

SUMMARY

Students are taking more complex pathways to earning STEM credentials than is generally assumed. They are likely to earn credits from more than one institution, to earn credits at a community college, and to transfer among institutions.

STEM students are also different than they were 25 years ago. The students are increasingly more likely to be from a minority group and to be single parents. The characteristics of students vary greatly across STEM disciplines, with rates of minority and female participation lowest in computer science, physics, and engineering.

The completion rates for students who aspire to a STEM degree remain lower than in non-STEM fields. At both 2-year and 4-year institutions, completion rates are lower for students from underrepresented groups compared to their white and Asian counterparts. Many students also take longer than expected to complete their credential. In addition, the goals of STEM aspirants (e.g., earning a degree or certificate, transferring to a 4-year institution, or gaining a specific job skill) and student populations vary across 2-year and 4-year institutions. Thus, it seems important to consider multiple factors (e.g., student goals, course completion, credit accumulation, time to and credits to degree, retention and transfer rates, degrees awarded, range of access) along with graduation rates when assessing the success of an institution.

The potential reasons for the low completion rates and differential rates across groups are explored in the following chapters.

REFERENCES

American Society for Engineering Education. (2012). *Innovation with Impact: Creating a Culture for Scholarly and Systematic Innovation in Engineering Education.* Washington, DC: American Society for Engineering Education.

Atman, C.S.D. Sheppard, J., Turns, R.S., Adams, L.N., Fleming, R., Stevens, R.A., Streveler, R., Smith, K.A., Miller, R.L., Leifer, L.J., Yasuhara, K., and Lund, D. (2010). *Enabling Engineering Student Success: The Final Report for the Center for the Advancement of Engineering Education.* Available: http://www.engr.washington.edu/caee/CAEE%20 final%20report%2020101102.pdf [July 2015].

Brainard, S.G., and Carlin, L. (1998). A six-year longitudinal study of undergraduate women in engineering and science. *Journal of Engineering Education, 87*(4), 369–375.

Crosta, P.M. (2014). Intensity and attachment: How the chaotic enrollment patterns of community college students relate to education outcomes. *Community College Review, 43,* 46–71.

The College Board. (2014). *Trends in College Pricing 2013.* Available: https://trends. collegeboard.org/sites/default/files/college-pricing-2013-full-report.pdf [April 2015].

Complete College America. (2014). *Four-Year Myth: Make College More Affordable, Restore the Promise of Graduating on Time.* Indianapolis, IN: Complete College America.

Dym, C.L., Agogina, A.M., Eris, O., Frey, D.D., and Leifer, L.J. (2005). Engineering design thinking, teaching, and learning. *Journal of Engineering Education, 94*(1), 103–120.

Eagan, M.K., Hurtado, S., Chang, M.J., Garcia, G.A., Herrera, F.A., and Garibay, J.C. (2013). Making a difference in science education: The impact of undergraduate research programs. *American Educational Research Journal, 50*(4), 683–713.

Eagan, K., Hurtado, S, Figueroa, T., and Hughes, B. (2014). *Examining STEM Pathways among Students Who Begin College at Four-Year Institutions.* Commissioned paper prepared for the Committee on Barriers and Opportunities in Completing 2- and 4-Year STEM Degrees, National Academy of Sciences, Washington, DC. Available: http://sites. nationalacademies.org/cs/groups/dbassesite/documents/webpage/dbasse_088834.pdf [April 2015].

Freeman, S., Eddy. S.L., McDonough, M., Smith, M.K., Okoroafor, N., Jordt, H., and Wenderoth, M.P. (2014). End of lecture: Active learning increases student performance across the STEM disciplines. *Proceedings of the National Academy of Sciences of the United States of America, 111*(23), 8410–8415.

Goldrick-Rab, S. (2006). Following their every move: How social class shapes postsecondary pathways. *Sociology of Education, 79*(1), 61–79.

Hoit, M., and Ohland, M. (1998). The impact of a discipline-based introduction to engineering course on improving retention. *Journal of Engineering Education, 87*(1), 80–85.

Hossler, D., Shapiro, D., Dundar, A., Ziskin, M., Chen, J., Zerquera, D., and Torres, V. (2012). *Transfer and Mobility: A National View of Pre-Degree Student Movement in Postsecondary Institutions.* Herndon, VA: National Student Clearinghouse Research Center. Available: http://pas.indiana.edu/pdf/Transfer%20&%20Mobility.pdf [April 2015].

Hurtado, S., Eagan, M.K., and Hughes, B. (2012). *Priming the Pump or the Sieve: Institutional Contexts and URM STEM Degree Attainments.* Paper presented at the Annual Forum of the Association for Institutional Research. New Orleans, LA. Available: http://www.heri. ucla.edu/nih/downloads/AIR2012HurtadoPrimingthePump.pdf [October 2015].

The Institute for College Access and Success. (2014). *Quick Facts about Student Debt.* Available: http://bit.ly/1lxjskr [April 2015].

Kena, G., Aud, S., Johnson, F., Wang, X., Zhang, J., Rathbun, A., Wildinson-Fliker, S., and Kristapovich, P. (2014). *The Condition of Education.* Washington, DC: U.S. Department of Education, National Center for Education Statistics.

Kinser, K. (2014). *For-Profit Pathways into STEM*. Presentation to the Committee on Barriers and Opportunities in Completing 2-Year and 4-Year STEM Degrees. Available: http://http://sites.nationalacademies.org/cs/groups/dbassesite/documents/webpage/dbasse_086697.pdf [April 2015].

Lichtenstein, G., Loshbaugh, H.G., Claar, B., Chen, H.L., Jackson, K., and Sheppard, S. (2009). An engineering degree does not (necessarily) an engineer make: Career decision making among undergraduate engineering majors. *Journal of Engineering Education, 98*(3), 227–234.

Lichtenstein, G., Chen, H.L., Smith, K.A., and Maldonado, T.A. (2014). Retention and persistence of women and minorities along the engineering pathway in the United States. In A. Johri and B.M. Olds (Eds.), *Cambridge Handbook of Engineering Education Research*. New York: Cambridge University Press.

Mooney, G.M., and Foley, D.J. (2011). *Community Colleges: Playing an Important Role in the Education of Science, Engineering, and Health Graduates*. Arlington, VA: National Science Foundation.

National Academy of Engineering. (2004). *The Engineer of 2020: Visions of Engineering in the New Century*. Washington, DC: The National Academies Press.

National Academy of Engineering. (2005). *Educating the Engineer of 2020: Adapting Engineering Education to the New Century*. Committee on the Engineer of 2020, Phase II, Committee on Engineering Education. Washington, DC: The National Academies Press.

National Academy of Engineering. (2008). *Changing the Conversation: Messages for Improving Public Understanding of Engineering*. Committee on Public Understanding of Engineering Messages. Washington, DC: The National Academies Press.

National Academy of Engineering and American Society for Engineering Education. (2014). *Surmounting the Barriers: Ethnic Diversity in Engineering Education. Summary of a Workshop*. Washington, DC: The National Academies Press.

National Academy of Sciences, National Academy of Engineering, and Institute of Medicine. (2007). *Rising Above the Gathering Storm: Energizing and Employing America for a Brighter Economic Future*. Committee on Prospering in the Global Economy of the 21st Century: An Agenda for American Science and Technology. Committee on Science, Engineering, and Public Policy. Washington, DC: The National Academies Press.

National Center for Education Statistics. (1991). *Digest of Education Statistics 1990*. Washington, DC: U.S. Department of Education.

National Center for Education Statistics. (2012). *Integrated Postsecondary Education Data System*. Washington, DC: U.S. Department of Education.

National Center for Education Statistics. (2014). *Digest of Education Statistics 2014*. Washington, DC: U.S. Department of Education.

National Center for Education Statistics. (2015). *Digest of Education Statistics 2015*. Washington, DC: U.S. Department of Education.

National Research Council. (2011). *Expanding Underrepresented Minority Participation: America's Science and Technology Talent at the Crossroads*. Committee on Underrepresented Groups and Expansion of the Science and Engineering Workforce Pipeline, F.A. Hrabowski, P.H. Henderson, and E. Psalmonds (Eds.). Board on Higher Education and the Workforce, Division on Policy and Global Affairs. Washington, DC: The National Academies Press.

National Science Board. (2014). *Science and Engineering Indicators 2014*. NSB 14-01. Arlington VA: National Science Foundation.

National Student Clearinghouse Research Center. (2012). *The Role of 2-Year Institutions in 4-Year Success: Snapshot Report*. Herndon, VA: National Student Clearinghouse Research Center. Available: http://nscresearchcenter.org/wp-content/uploads/SnapshotReport6-TwoYearContributions.pdf [October 2015].

Ohland, M.W., Sheppard, S.D., Lichtenstein, G., Eris, O., and Chachra, D. (2008). Persistence, engagement, and migration in engineering programs. *Journal of Engineering Education,* 7(3), 259–278.

President's Council of Advisors on Science and Technology. (2012). *Report to the President Engage to Excel: Producing One Million Additional College Graduates with Degrees in Science, Technology, Engineering and Mathematics.* Available: http://www.whitehouse. gov/sites/default/files/microsites/ostp/pcast-engage-to-excel-final_feb.pdf [April 2015].

Ramsey, L.R., Betz, D.E., and Sekaquaptewa, D. (2013). The effects of an academic environment intervention on science identification among women in STEM. *Social Psychology of Education,* 16(3), 377–397.

Rassen, E., Chaplot P., Jenkins, D.P., and Johnstone, R. (2013). *Principles of Redesign: Promising Approaches to Transforming Student Outcomes.* New York: Community College Research Center, Teachers College, Columbia University.

Russell, J.S., and Lenox, T.A. (2013). *Raise the Bar: Strengthening the Civil Engineering Profession.* Reston, VA: American Society of Civil Engineers.

Ruch, R.S. (2001). *Higher Ed., Inc.: The Rise of the For-Profit University.* Baltimore, MD: Johns Hopkins University Press.

Salzman, H., and Van Noy, M. (2014). *Crossing the Boundaries: STEM Students in Four-Year and Community Colleges.* Commissioned paper prepared for the Committee on Barriers and Opportunities in Completing 2-Year and 4-Year STEM Degrees, National Academy of Sciences, Washington, DC. Available: http://sites.nationalacademies.org/cs/groups/dbasse site/documents/webpage/dbasse_089924.pdf [April 2015].

Sax, L.J., Hartado, S., Lindholm, J.A., Astin, A.W., Korn, W.S., and Mahoney, K.M. (2004). *The American Freshman: National Norms for Fall 2004.* Los Angeles: Higher Education Research Institute, University of California.

Seymour, E., and Hewitt, N. (1997). *Talking about Leaving: Why Undergraduates Leave the Sciences.* Boulder, CO: Westview Press.

Sheppard, S.D., Macatangay, K., Colby, A., and Sullivan, W.M. (2008). *Educating Engineers: Designing for the Future of the Field.* San Francisco: Jossey Bass.

Sheppard, S.D., Antonio, A., Brunhaver, S., and Gilmartin, S. (2014). Studying the career pathways of engineers: An illustration with two datasets. In A. Johri and B. Olds (Eds.), *Cambridge Handbook of Engineering Education Research* (pp. 283–310). Cambridge, MA: Cambridge University Press.

Smith, K.A., Sheppard, S.D., Johnson, D.W., and Johnson. R.T. (2005). Pedagogies of engagement: Classroom-based practices (cooperative learning and problem-based learning). *Journal of Engineering Education,* 94(1), 87–101.

Van Noy, M., and Zeidenberg, M. (2014). *Hidden STEM Knowledge Producers: Community Colleges' Multiple Contributions to STEM Education and Workforce Development.* Commissioned paper prepared for the Committee on Barriers and Opportunities in Completing 2-Year and 4-Year STEM Degrees, National Academy of Sciences, Washington, DC. Available: http://sites.nationalacademies.org/cs/groups/dbassesite/documents/ webpage/dbasse_088831.pdf [April 2015].

Wang, X. (2013). Why students choose STEM Majors: Motivation, high school learning, and postsecondary context of support. *American Educational Research Journal,* 50(5). 1081–1121.

3

The Culture of Undergraduate
STEM Education

Major Messages

- The culture of science, technology, engineering, and mathematics (STEM) education has an effect on many students' interest, self-concept, sense of connectedness, and persistence in these disciplines.
- New research is needed to understand whether STEM "gateway" courses continue to negatively impact STEM student persistence due to the culture of the classrooms and a heavy reliance on lectures, as research from over a decade ago has revealed.

The complex array of pathways that students take to STEM degrees is not easily navigated, and students sometimes encounter barriers along the path to earning a degree. The environments they encounter when they begin college may not be welcoming, and the teaching may be uninspired. Barriers also result from departmental, institutional, and national policies. They may find themselves inadequately prepared for the rigor of college coursework or they may face stereotypes from faculty or peers. Students may encounter these barriers in classrooms and in other aspects of campus life. In this chapter, we address the barriers that students encounter related to the culture of STEM education: that is, the shared patterns of norms, behaviors, and values of STEM disciplines that manifest themselves in the way courses are taught and the classroom is experienced. We explore barriers related to instructional quality and policy barriers in the following chapters. By "culture," we mean the explicit and implicit customs and behaviors, norms, and values that are normative within STEM education (National Research Council, 2009). It is important to focus on the culture of STEM education because the social, psychological, and structural dimensions of STEM education in colleges and universities influence how students connect their personal identities to their academic domains and view themselves as learners in those domains (their academic identities), which subsequently affects their efforts and achievement (Cabrera et al., 1999; Eccles et al., 1998; Reid and Radhakrishnan, 2003; Perez et al., 2014). The academic climate that individual students experience in college—their perceptions of interpersonal interactions and norms—is a manifestation of the college culture and one factor that influences student performance, engagement, and persistence outside of what would be predicted by socioeconomic or academic preparation indicators (Chang et al., 2011).

The importance of culture cuts across all institution types and pathways to STEM credentials. College campuses and the STEM departments and programs in them represent distinct types of organizational settings, with cultures created and perpetuated by physical structures, policies, underlying values, and social norms that guide their functioning. The cultures that students experience shape their awareness and understanding of standards, expectations, and their belonging. For example, the small numbers and limited examples of black professionals in such fields as geosciences might lead to perceptions by those in the field that reinforce the belief that "black people don't do geology." Similarly, in traditionally male-dominated professions, such as engineering, women may need to overcome explicit and subtle cultural messages that men are better suited for such professions (Cech and Waidzunas, 2011). The cultures that male and female students from all backgrounds, races, and ethnicities encounter while they study STEM can undermine or support their performance and persistence through their self-concepts and beliefs specific to the STEM domain and their feelings of community and belonging in STEM fields. In this chapter we focus on how the culture of STEM education impacts women and underrepresented students because of the concerns about participation of students from these groups in STEM fields and because students from these groups are typically the subject of research on the effect of the culture of STEM education.

The relationship between institutional or disciplinary culture and race, ethnicity, and gender is especially relevant in STEM fields, where racial and ethnic minorities and women are even more underrepresented than they are in most other fields (Anderson et al., 2006; National Research Council, 2011). For historically underrepresented students, views of the

BOX 3-1
The Value of Diversity

Perspective may be an important aspect of problem solving in science. . . . What is considered creativity on the part of an individual may in fact be a different perspective. In order to solve problems which are currently considered intractable, it may be critical to involve people who are traditionally not participants in the scientific process, especially women.

Induction into Western New York Women's Hall of Fame, 2011
Esther S. Takeuchi

As described in *Expanding Underrepresented Minority Participation: America's Science and Technology Talent at the Crossroads* (National Academy of Sciences, National Academy of Engineering, and Institute of Medicine, 2011), diversity is a resource for and strength of the nation's society, economy, and postsecondary institutions. Diverse groups are typically smarter and stronger than homogeneous groups when innovation is a critical goal (Page, 2007). Greater diversity in an institution, therefore, strengthens it by increasing the number of perspectives and the range of knowledge represented.

Diversity initiatives positively affect both minority and majority students on campus in terms of student attitudes toward racial issues, institutional satisfaction, and academic growth (Smith, 1997). Diversity in disciplinary work contributes to the research agendas of individual faculty and their departments, aligns with scholarly values, and promotes such student learning goals as tolerance of ambiguity and paradox, critical thinking, and creativity (Anderson, 2008).

Work by Gurin and colleagues (2002) illustrate three key benefits of diversity in postsecondary education. First, structural diversity creates conditions that lead students to experience diversity in ways that would not occur in a more homogeneous student body. Second, students who experience the most diversity in classroom settings and in informal interactions with peers show the greatest engagement in active thinking processes, growth motivation, and growth in intellectual and academic skills. Third, higher education plays a central role in helping students to become active citizens and participants in a pluralistic democracy. Students who experience diversity in classroom settings and in informal interactions show the most engagement in various forms of citizenship and the most engagement with people from different races and cultures.

way race, ethnicity, and gender function in their college environment are especially important in their social and academic adjustment (Reid and Radhakrishnan, 2003). Experiencing a college culture with a hostile or unwelcoming racial environment has been related to social and academic withdrawal (Cabrera et al., 1999; Hurtado et al., 1998; Yosso et al., 2009), academic and social isolation (Allen, 1988; Fleming, 1984; Nettles, 1988;

Ali and Kohun, 2006; Strayhorn, 2010a, 2012), and a host of other negative consequences (see below). In situations where students are underrepresented—as the only woman or Hispanic person in a class or department, for example—their social identities are more salient to both minority and majority group members (Hurtado et al., 1996). The value of cultivating diversity in science is described in Box 3-1.

WAYS OF KNOWING AND DISCOURSE IN STEM EDUCATION

As described in a previous National Research Council (2009) study, conceiving of culture as shared repertoires of practices sometimes leads researchers to refer to membership in almost any type of group as membership in a culture. This conceptualization of culture is highly relevant to undergraduate STEM education, which prepares students to become members of a group: professional scientists, technologists, engineers, or mathematicians. Thus, STEM learning can be viewed as a cultural process in which the practices and assumptions of STEM education reflect the culture, cultural practices, and cultural values of STEM professionals (National Research Council, 2009). From this perspective it is not surprising to find that a STEM educator's notion of what counts as scientific reasoning and sense-making practices reflects those that are valued and used by STEM professionals (Ballenger, 1997).

An educator's notion of what counts as scientific reasoning and sense-making can become a barrier for some STEM aspirants. For example, the discursive norms in STEM classrooms around debate and argumentation with student peers and instructors may not reflect students' own prior experiences and norms in their communities and schools (Brown, 2004; Kurth et al., 2002). An example is the idea of argumentation with an elder, which is not seen as acceptable behavior in some communities. Similarly, researchers have characterized the language of STEM as reflecting white, middle-class, masculine norms, which may be at odds with norms of expression more likely found among women and students from historically underrepresented groups (Brandt, 2008; Lemke, 2001; Olitsky, 2006); this disconnect can prevent them from identifying with STEM (Carleone and Johnson, 2007; Olitsky, 2006; Ong, 2005).

In other cases, students must first recognize and then negotiate and reconcile differences between their culturally based epistemological beliefs and those of mainstream science contexts, which may be invisible to instructors or be perceived as resistance or disengagement (Nelson-Barber and Estrin, 1995). This barrier is particularly salient for Native Americans and Alaska Natives, whose ways of knowing and views of the natural world often diverge from those present in STEM classrooms (Aikenhead, 1998; Bang et. al., 2007; Cobern and Aikenhead, 1998). Native American and Alaska Native students may be marginalized by STEM instruction that portrays scientific ways of knowing as free from value and above the influence of context, because such instruction is at odds with their cultural self-identity (Aikenhead and Ogawa, 2007). In fact, Aikenhead (2001) argues that only a small minority of students have world views and self-identities that align with the ways of knowing frequently conveyed in STEM classrooms.

A barrier that many students experience within the normative culture of STEM includes the view that inherent or natural ability determines a person's capacity for STEM learning, more so than other subject domains (Crisp et al., 2009; Dai and Cromley, 2014; Smith et al., 2013). A belief that natural ability determines capacity for STEM may vary by field. Recent research has shown that the extent to which professionals in STEM fields believe that innate talent is required for success is a strong predictor of representation of women and blacks in that field (Leslie et al., 2015). Fields where professionals believe innate talent is necessary tend to have fewer women and minorities. The overall message conveyed is that success in STEM fields requires either natural ability in mathematics or science or very early exposures to high-quality training. Related to this view is the tendency for introductory mathematics and science courses to function as gatekeeper courses that discourage students from continuing to pursue a STEM degree: see Box 3-2 for a detailed discussion of mathematics. Although practices and structures may vary across institutions and STEM departments, there are concerns that STEM gateway courses are characterized by a culture of highly competitive classrooms that do not promote active participation. The implied goal of these courses is to distinguish between those believed to have the ability to succeed in STEM from those who do not and "select out" the latter (Crisp et al., 2009; President's Council of Advisors on Science and Technology, 2012). In such settings, students from historically underrepresented backgrounds may be particularly likely to experience low expectations exacerbated by bias and small numbers of students from their group (their token status) in the field. Empirical support for these concerns is limited to a small number of studies with a limited sample and data from the mid-1990s (Gainen, 1995; Seymour and Hewitt, 1997). Additional studies of the nature of instructional strategies and the classroom culture are needed to determine if the continued criticism of these courses is warranted.

BOX 3-2
Mathematics

The mathematical sciences are unique among the STEM disciplines in that they serve both as important and evolving fields of advanced study and as a source of foundational knowledge required of every STEM major. Whether or not a student studied calculus in high school is one of the strongest predictors of successful completion of a STEM degree (Chen, 2009). Sixty percent of those who declare a STEM major and studied calculus in high school will complete their STEM degree. In addition, while AP biology, chemistry, and physics give students an advantage in the particular discipline—biology, chemistry, or physics—only AP Calculus has an effect that transfers to other STEM disciplines (Sadler and Tai, 2007). Mathematics instruction significantly affects student learning in STEM with respect to how students create, use, interpret, and translate graphical and mathematical representations in the context of conceptual understanding in a discipline. Although much could be noted about the challenges and opportunities of introducing students to the potential of a major in mathematics, we focus here on the issues facing mathematics as the gateway to other STEM disciplines.

For many students interested in earning a STEM credential, their mathematical performance in high school has been very high, but for some it has not. Each group faces its own particular set of challenges and obstacles, in addition to the obstacles that frequently face all students including class size, the need for curricular coherence, the nature of the pedagogy, and the availability of student support services.

Students who performed well in high school mathematics still struggle in calculus I. Twenty-five percent of the students who take calculus I at a research university receive a D or F or withdraw from the course (DFW), and another 23 percent receive a C (Bressoud et al., 2013), a grade that is widely perceived as a signal that one is not adequately prepared to succeed in calculus II (Tyson, 2011). Perhaps the most striking finding about high-performing students is the tremendous loss of confidence in their first term of university-level mathematics. This phenomenon is particularly strong for women. Women who complete calculus I with a grade of A or B are less likely than men to continue on to calculus II (Rasmussen and Ellis, 2013). It is worth noting that no differences by race or ethnicity were observed among high-performing students of the same gender (Tyson, 2011; Rasmussen and Ellis, 2013).

Lower-performing, but not necessarily low-performing, students often take their first college mathematics course at a community college (Bressoud, 2014). These students face the obstacle of college algebra or precalculus, as well as the need to take precollege-level (noncredit) mathematics courses. All of these courses are notoriously ineffective at advancing students to the level needed for success in calculus. First, good placement procedures are rare (Carlson et al., 2010). Second, even those who do well in precalculus (C or higher) often do not go on to enroll in calculus (Thompson et al., 2007; Herriott and Dunbar, 2009). This trend is particularly pronounced among students intending to major in STEM fields. For example, a study of students at public and private institutions found that only half of the students intending to major in STEM who took precalculus enrolled in calculus I and only 40 percent of them eventually enrolled in calculus II (Herriott and Dunbar, 2009). Moreover, taking precalculus prior to calculus I seems to have little if any effect on student performance in calculus I (Hsu et al., 2008; Sonnert and Sadler, 2014).

Most students who were low performing in high school mathematics and seek a STEM degree start at a 2-year college. These students face a long and difficult succession of courses that must be negotiated before they can take calculus. The succession often begins with precollege level or developmental mathematics. Only about 30 percent of students successfully complete developmental mathematics and only 20 percent of those who complete the developmental course go on to complete a college-level mathematics course (Bailey et al., 2010). Thus, developmental mathematics courses, particularly in the context of community colleges, are a barrier to student success in undergraduate STEM education. They are a barrier to success because the courses add time and cost to degree completion, while they do not succeed in preparing the majority of students for college-level mathematics.

A growing number of strategies have been developed to improve undergraduate mathematics education. The use of technology is a key aspect of one set of strategies. An example in precalculus is Assessment and Learning in Knowledge Spaces (ALEKS). It is an adaptive testing platform that can be used to evaluate student knowledge of fine-grained topics up to and including precalculus. Building a precalculus course around ALEKS has proven to be very successful at several universities. One example is the University of Illinois at Urbana–Champaign (Ahlgren and Harper, 2011; Hagerty et al., 2005, 2010). ALEKS has several advantages: one is that exams are individualized and offered online so that students can take them where and when they so wish, and adaptive testing means that there is an opportunity to drill into particular competencies to assess exactly what students can and cannot do. In addition, some studies have found that intelligent tutoring systems, such as Cognitive Tutors, and the Open Learning Initiative, have been shown to have positive effects on college students' understanding of mathematics, performance in mathematics courses, and persistence in college (Koedinger and Sueker, 1996; Scheines et al., 2005; Bowen et al., 2012; Kaufman et al., 2013; Ritter, 2014). However, other studies have not found positive effects of intelligent tutoring systems (Campuzano et al., 2009; Pane et al., 2010). The variation in effect seems to be due to whether the tutors were implemented with fidelity (Pane et al., 2010). Implementing intelligent tutoring systems as part of a blended classroom (i.e., instruction that is delivered in the classroom and through digital media) has been shown to lead to better academic outcomes, over a shorter period of time, than traditional courses (Bowen et al., 2012).

Another strategy to improve undergraduate mathematics education has been to revise the pathway to calculus I. For example, many postsecondary institutions now offer stretched-out versions of calculus I (Bressoud, 2014). Such a course, spread over two terms, embeds review of precalculus topics on a just-in-time basis. This approach combines new and challenging mathematics with the opportunity to review and reintroduce areas of weakness. Materials for such a course were first developed at Moravian College in the 1990s (Sevilla and Somers, 1993). The Wright State University model for engineering programs has seen increases in student retention by delaying the calculus portion of the curriculum until after students have taken introductory engineering courses with embedded math (Klingbeil et al., 2006). The Community College Pathways, the California Acceleration Project, and New Mathways Project have demonstrated success in improving undergraduate mathematics education by altering the sequence of mathematics courses that low-performing students take and adjusting the instructional methods within mathematics courses.

BELIEFS ABOUT ABILITY TO LEARN STEM

Increasingly, studies of college students have linked students' beliefs about their academic ability in STEM to their STEM performance and persistence (Carleone and Johnson, 2007; Chemers et al., 2011; Perez et al., 2014; Williams and George-Jackson, 2014). Emerging research illustrates how negative ability cues and stereotypes in college can be overcome.

Ability cues (signals of what ability is, who has it, and who does not) are commonly conveyed in academic settings and are embedded in their structures and practices. These cues can influence students' views of their own ability. Research on implicit beliefs about ability show that students who think of ability as fixed respond to academic settings in different ways than those who think of ability as malleable (see, e.g., Dweck and Leggett, 1988). Students with fixed beliefs about ability are more likely to avoid challenging tasks and to view challenge as more threatening to their self-concepts. They are more likely to respond to challenge or failure by feeling helpless, avoiding help-seeking, and ultimately, disengaging. In contrast, students who view ability as malleable view failures as opportunities to learn, are persistent in the context of challenge or failure, and are more likely to seek help (Dweck, 2000). Thus, believing that ability in STEM can improve with learning and effort is related to positive motivational responses and performance outcomes (Dai and Cromley, 2014). In fact, Dai and Cromley (2014) have shown that increases in fixed beliefs following entry into STEM courses predicted dropout in biology, beyond a student's grade. The increases in fixed beliefs were found to be associated with messages conveyed in gateway courses. The authors argue that the structure of the curriculum and instructional strategies are associated with changes in students' mindsets, thus, leading to engagement (with decreases in fixed beliefs) or disengagement (with increases in fixed beliefs).

Multiple studies have shown significant positive effects of interventions that target students' beliefs about their ability to succeed in STEM by suggesting that the causes of low grades are unstable (i.e., related to effort rather than ability) (reviewed in Snipes et al., 2012). For example, in an intervention developed by Wilson and Linville (1985), some struggling first-year college students were shown videos of college seniors discussing how their grades were low in their first year but had improved over time through hard work (Snipes et al., 2012). There is evidence from a number of studies that students who were randomly assigned to such interventions do better on both short-term and long-term performance measures. While there are a number of promising interventions and tools, there are questions regarding how to take the interventions and tools to scale. In particular, more research is needed to flesh out the interactions among target populations, educational contexts, and instructional strategies (Snipes et al., 2012).

Negative race and gender stereotypes about ability are particularly salient in STEM fields and may convey signals around the inherent or fixed nature of ability. For instance, research has noted the "undervaluing" of females and minorities in STEM, with lower expectation of their presence among geniuses (Hyde and Mertz, 2009). Thus, in STEM fields, underrepresented minorities and women may be particularly vulnerable to disengagement (leaving a STEM field of study) due to beliefs about their ability to succeed in STEM, even when accounting for prior academic preparation (Litzler et al., 2014).

These common stereotypes can be overcome, however. In one study (Aronson et al., 2002), students who received explicit messages in classroom settings around the incremental nature of ability (that it can improve over time with instruction and practice) at the beginning of their academic term showed greater academic enjoyment and engagement and higher performance at term's end than did students who did not receive such instruction. The positive relationship between messages and students' outcomes was observed for all students in the study with the strongest effects among black students. Thus, academic climates that emphasize learning, mastery, and improvement in math and science, rather than inherent ability, can promote both performance and persistence in those subjects through positive effects on students' self-beliefs. This effect may be especially strong for historically underrepresented groups, for whom negative academic stereotypes may be present in both subtle and overt ways in their day-to-day academic lives.

COMMUNITY BELONGING AND STEM EDUCATION

In addition to self-beliefs, students' connections to their campus communities can enhance their academic engagement and, subsequently, students' identification with their discipline, including their positive affect (feelings) toward the discipline (Fleming, 1984; Good, 2012; Hurtado et al., 2008; Johnson, 2011, 2012; Ko et al., 2014; Locks et al., 2008; London et al., 2014; Palmer et al., 2011). Connection to community covers both a sense of belonging to an academic setting (an institution, a department, or subgroups within them) and a psychological sense of community (a broader connection to the discipline or field area).

One study of an introductory electrical engineering class at a major university in the Northwest (Lee et al., 2006) found that positive affect and positive relationships with others were correlated with positive classroom experiences. The study also found that students with positive classroom experiences had a more positive career outlook. In contrast, students who do not experience a sense of community or belonging in STEM fields are more likely to leave STEM majors (Smith et al., 2013). Women report that "isolation" is a primary reason for their choice to leave science, technology, and engineering (Brainard and Carlin, 1998; Hewlett et al., 2008). Also, women's ambivalence about their belonging in computer science has been linked to their low level of participation in the field (Cheryan et al., 2009; Wolcott, 2001).

RACIAL AND GENDER STEREOTYPES AND BIASES IN STEM EDUCATION

A host of psychological and educational research studies provides clear evidence that stigmatizing experiences—in the form of interpersonal discrimination—are a common occurrence for many racial and ethnic minority students, especially those in predominantly white college and university settings (see, e.g., Chang et al., 2011). This also occurs for women in STEM fields in which they are underrepresented (Brainard and Carlin, 1998; Hughes, 2012; Ramsey et al., 2013; Reyes, 2011). These experiences are a source of educational inequity as they negatively affect the quality of many of these students' social and academic experiences (Chang et al., 2011). These negative experiences can lead to a decreased sense of connectedness and community within students' academic settings.

When individuals perceive that negative stereotypes about their group are salient in a particular situation or context, they experience "stereotype threat" (Steele, 1997). The "threat" is represented by individuals' apprehension that they may be viewed in ways that are consistent with group stereotypes. Numerous studies have demonstrated that stereotype threat negatively affects performance on academic tasks (e.g., Aronson and Salinas, 1997; Gonzales et al., 2002; McKay et al., 2002; Schmader and Johns, 2003; Steele and Aronson, 1995). Under repeated stereotype threat conditions (i.e., typical day-to-day academic contexts in which stereotypes are often salient), students may respond by psychologically disconnecting their personal identity from the academic domain (academic dis-identification). In doing so, students may come to minimize attributes and behaviors necessary for success in their educational domain and develop personal identities in areas outside of that domain (Cokley, 2000; Crocker and Major, 1989; Osborne, 1997, 1999; Steele, 1997). Although this coping response may help protect students' self-concept, it undermines the motivation and engagement necessary for positive performance and persistence in an academic domain. For instance, it is possible that repeated exposure to stereotype threat in STEM courses among underrepresented students who intend to earn a STEM degree leads these students to "dis-identify" with STEM while at the same time retaining their connections to education and college more generally. In doing so, they still may be successful in attaining a college degree in another major area, but they would be less likely to attain STEM degrees or aspire to pursue STEM graduate degrees or careers.

In addition to indirect messages about ability and belonging embedded in academic cultures in higher education, there is evidence that underrepresented students—relative to their majority peers—commonly encounter more overt stigma experiences. Those experiences have been characterized as microaggressions, from instructors, peers, administrative staff, and other staff. These microaggressions are subtle or overt statements and behaviors that intentionally or unintentionally communicate devaluing messages about a group, including expressed low expectations (e.g., Fries-Britt and Griffin, 2007; Hurtado et al., 2011; Nadal et al., 2014; Solorzano et al., 2000; Yosso et al., 2009). Experiences of microaggressions can lead to feelings of invisibility: students feel as if they are viewed only in terms of stereotypes rather than in terms of their unique identities and characteristics (Franklin and Boyd-Franklin, 2000). For example, in one study (Smith et al., 2011), black male students who experienced stereotype-based treatment in their daily college contexts (e.g., being treated as intellectually inferior or as criminally deviant) were more likely to have feelings of isolation on campus that inhibited their academic performance.

An equally insidious phenomenon is the "benign racism" or "benign sexism" that can occur in mentoring, referred to as the "mentor's dilemma" (Cohen et al., 1999). This dilemma refers to faculty who are mentoring students across "cultural lines." In such cases, faculty members are less likely to provide tough, specific feedback to minority students due to concerns about appearing biased. Instead, faculty members may overpraise performance or effort and provide vague feedback in attempts to affirm students and convey a supportive environment or to "protect" students' self-esteem. Often these actions reflect faculty's implicit biases, based in negative cultural stereotypes about ability. Consequently, underrepresented students do not access and benefit from the same high-quality feedback as do other students. In both cases, students' experiences signal perceptions of their low ability in ways that can undermine their self-concept and subsequent engagement. Unfortunately, these experiences can serve to undermine students' own views of their ability and make them feel less valued, and subsequently, less connected to their academic settings. The experiences of female students from underrepresented minority groups are discussed in Box 3-3.

Changes to departmental or institutional culture can make a difference. A recent study (Ramsey et al., 2013) compared women in STEM departments characterized by welcoming versus traditional (unwelcoming) cultures for women. The welcoming cultures were characterized by more positive, visible messages about women in STEM, more women identifying in STEM in visible ways (carrying or wearing markers of STEM majors), and more peer role models. The women in the welcoming climate had fewer

BOX 3-3
The Double Bind Effect

The challenges that minority women face in pursuing a science, technology, engineering, and mathematics (STEM) degree were first called out in *The Double Bind: The Price of Being a Minority Woman in Science* (Malcom et al., 1976). A follow-up paper considered the situation 35 years later (Malcom and Malcom, 2011, p. 162):

> [M]uch has changed, and much has not changed. STEM fields continue to be overwhelmingly dominated by Whites and men, although the passage of laws banning discrimination on the basis of race and/or sex reduced the number of overt practices that shaped the university and workforce cohorts of previous years.

The barriers faced by minority women today are not as overt as the discriminatory practices and policies of 35 years ago: they are more related to a lack of support and inaction by institutions. Minority men also face barriers in their pursuit of a STEM degree (see National Academy of Engineering, 2012). We focus on minority women in order to provide a rich discussion of the barriers encountered as a result of the interaction between race and gender.

Both minority women and women in general are more heavily concentrated in community colleges, for-profit institutions, and less-selective colleges and universities than their white and male counterparts. Minority women represent a disproportionate number of students who received an associate's degree at a community college prior to earning a STEM baccalaureate degree (Malcom et al., 2010).

The potential contribution of minority women is significant given that they express strong interest in STEM fields and greater intention to major in these fields in postsecondary study than do white females (National Science Foundation, 2013; Riegle-Crumb and King, 2010). However, minority women face many institutional and cultural barriers to achieving their goal of completing a STEM degree.

A synthesis of 116 works of scholarship spanning 40 years (Ong et al., 2011) provides insight into the factors that influence the retention, persistence, and achievement of underrepresented minority women in STEM fields. Those complex and interrelated factors include personal relationships (faculty, peers, and family), STEM enrichment programs, sense of academic self, individual agency and drive, and the climate of the learning environment.

Underrepresented minority women have to do a "tremendous amount of extra, and indeed, invisible work" (Ong, 2002, p. 43) in order to fit in with and gain the respect of the white male physics peers and faculty. In addition, this study found that the women had to spend more effort learning the unspoken rules of the culture of physics to gain and maintain "membership" in the culture. Yet another study (Ong et al., 2011) found that a supportive climate for underrepresented minority women, particularly at historically black colleges and universities, led to lower rates of attrition in STEM majors. Institutional factors that were correlated with lower attrition rates among underrepresented minority women included openness to alternative routes to a STEM major (i.e., a lack of stigma for remedial work), high expectations, and supportive relations between students and faculty.

A study of 1,250 underrepresented minority women and 891 white women at more than 130 institutions found that underrepresented minority women persisted in STEM degree programs at private colleges where there was a "robust community of STEM students" (Espinosa, 2011). One of the strongest negative predictors of persistence was attending a highly selective school.

Underrepresented minority women (excluding Asian/Pacific Islanders) account for 17 percent of the undergraduate STEM degrees awarded in 2012 (National Science Foundation, 2013). The women are heavily overrepresented in some fields and heavily underrepresented in others: they accounted for 25 percent of psychology degrees, 19 percent of social science degrees, and 15 percent of biology degrees: see Table 3-1 below. Together, these three majors accounted for about two-thirds of STEM degrees earned by underrepresented minority women, and they are underrepresented in all other STEM fields, especially computer science (7%), geosciences (7%), and engineering (4%).

TABLE 3-1 Science and Engineering Bachelor's Degrees Earned by Underrepresented Minority Women in 2012, by Field

| | Number of Degrees | | Percentage of Degrees | |
Field	All Students	Female	All Minorities	Minority Female
All Fields	1,810,647	57	28	17
Science and Engineering	589,330	51	27	15
All Sciences	506,067	56	28	17
Agricultural Sciences	25,060	54	16	9
Biological Sciences	99,900	59	23	15
Computer Sciences	47,960	18	30	7
Geosciences	5,865	39	14	7
Mathematics and Statistics	19,819	43	18	8
Physical Sciences	20,421	41	21	10
Psychology	109,716	77	32	25
Social Sciences	177,326	55	31	19
Engineering	83,263	19	18	4

NOTE: Underrepresented minority does not include Asian/Pacific Islanders.
SOURCE: Data from the National Science Foundation. Available: http://www.nsf.gov/statistics/wmpd/2013/pdf/tab5-7_updated_2014_05.pdf [April 2015].

concerns about whether they would succeed and increased STEM identification. This research demonstrated the potential for institutions to create (and re-create) STEM contexts in ways that enhance inclusion and participation for historically underrepresented groups.

CULTURAL STRENGTHS AND ASSETS OF STEM STUDENTS

Despite the risks and challenges faced by many minority students, significant numbers of STEM students from traditionally underrepresented groups show positive adjustment and are academically successful. There are a handful of systematic studies examining this within-group variation: How do personal, cultural, and contextual factors contribute to positive academic adjustment and to persistence in STEM fields of study? Several factors that may supersede or buffer the negative effects of stigmatizing contexts for many minority students have been identified (Chang et al., 2011), including parental support, intellectual development, and social connectedness to others, as well as students' awareness of and development of coping skills around experiences of racism and discrimination. Ko and colleagues (2014) illustrate that women's efforts to draw on personal, peer, and cultural supports are critical to maintaining their interest in science and their psychological well-being in contexts that devalue them. Women that persist in science often take extra strategic steps to get the mentoring and training they need when it is not provided in their academic settings (Ko et al., 2014). This study suggests that women and underrepresented students who persist in STEM may do so not necessarily because of changes or improvements in the STEM culture at their institution. Rather, they persist because of their agency and developed personal and cultural resources. Some of the co-curricular supports have been developed to support students' agency and personal and cultural resources (as discussed in Chapter 4).

Other studies have shown the need to acknowledge the strengths and positive educational values associated with minority students' cultural identities that are all too often ignored in favor of stereotypical views of minority groups as resisting and devaluing education (Hope et al., 2013; Ko et al., 2014; Yosso, 2005). For instance, scholars describe how black and Hispanic college students who experienced subtle and overt racism in their academic contexts actively pursued academic and professional excellence to "prove wrong" racialized and gendered assumptions and low expectations based in stereotypes (see, e.g., Fries-Britt and Griffin, 2007; Yosso et al., 2000, 2009).

Similarly, McGee and Martin (2011) examined the process of "stereotype management" among a sample of black college students in mathematics and engineering to explain their achievement and persistence. The students' moved from awareness that their racial identities were undervalued and feeling they needed to prove stereotypes wrong to their emphasizing the strengths associated with their racial and cultural identities, and to adopting more self-defined reasons to achieve. While this study found a connection between stereotype management and success in mathematics and engineering, students maintained a constant state of awareness that faculty and other students viewed black students as inferior in mathematics and engineering contexts. For example, as expressed in statements, such as "Really? Wow! I didn't think you would be able to answer a question like that! And no one helped you? (comment from an engineering professor directed to an African American female participant) (McGee and Martin, 2011, p. 2). In addition, the presence of stereotypes can be apparent to STEM students even when they are not expressed verbally or through nonverbal cues (McGee and Martin, 2011). As one student in the study explained, "even when no one uttered a word to him or gave him a 'What are you doing here?' glance, he still felt overwhelmed by the presence of that stereotype in most of his mathematics classrooms" (McGee and Martin, 2011, p. 18).

These lines of research challenge prevalent stereotypes and deficit perspectives of minority students as less able or less identified with academic pursuits. In addition, this research acknowledges student agency and avoids framing these students as passive victims of the types of unsupportive cultures and stigmatizing experiences they may face.

SUMMARY

The culture that students encounter when studying STEM has an effect on their interest, self-concept, sense of connectedness, and persistence in STEM. Many students encounter messages that success in STEM fields requires either natural ability in math or science or very early exposures to high-quality training, which tends to be associated with lower persistence among women and minorities. Academic cultures characterized by race, ethnic, or gender stigma may lead students to assess those academic contexts as incompatible with their personal identities; they may thus disidentify with or disconnect important aspects of their personal identity (e.g., self-esteem, self-concept, personal values) from the academic domain (Steele, 1992; Steele et al., 1998).

Students who persist often have to draw on personal, cultural, and cocurricular resources to counter messages about the nature of ability and stereotypes that they encounter in interactions with faculty and are embedded in organizational norms and practices. At the same time, institutions have the potential to create STEM academic climates that promote engagement, sense of connectedness, and persistence among students by positioning STEM as a context in which one can learn and develop, avoiding emphasis on inherent or natural ability. Institutions can also improve the academic climate by addressing the subtle and direct ways that students may experience messages and treatment in STEM contexts that are based on negative racial and gender stereotypes, including acknowledging and drawing on the cultural strengths that underrepresented students bring to their academic contexts and in efforts to develop or improve curricular and co-curricular practices and programs. These issues and others are discussed in detail in the following chapter.

REFERENCES

Ahlgren, A., and Harper, M. (2011). Assessment and placement through Calculus I at the University of Illinois. *Notices of the American Mathematical Society, 58*(10), 1460–1461.

Aikenhead, G. (1998). Many students cross cultural borders to learn science: Implications for teaching. *Australian Science Teachers' Journal, 44*(4), 9–12.

Aikenhead, G. (2001). Students' ease in crossing cultural borders into school science. *Science Learning, 85*(2), 180–188.

Aikenhead, G., and Ogawa, G. (2007). Ingenious knowledge and science revisited. *Cultural Studies of Science Education, 2*(3), 539–620.

Ali, A., and Kohun, F. (2006). Dealing with isolation feelings at IS doctoral programs. *International Journal of Doctoral Studies, 1*, 21–33.

Allen, W. (1988). The education of black students on white college campuses: What quality the experience? In M.T. Nettles (Ed.), *Toward Black Undergraduate Student Equality in American Higher Education* (pp. 57–86). New York: Greenwood Press.

Anderson, G.M., Sun, J.C., and Alfonso, M. (2006). Effectiveness of statewide articulation agreements on the probability of transfer: A preliminary policy analysis. *Review of Higher Education, 29*(3), 261–291. Available: http://muse.jhu.edu/login?auth=0&type=summary&url=/journals/review_of_higher_education/v029/29.3anderson.html [April 2015].

Anderson, J.A. (2008). *Driving Change Through Diversity and Globalization: Transformative Leadership in the Academy.* Sterling, VA: Stylus.

Aronson, J., Fried, C., and Good, C. (2002). Reducing the effects of stereotype threat on African American college students by shaping theories of intelligence. *Journal of Experimental Social Psychology, 38*, 113–125.

Aronson, J., and Salinas, M.F. (1997). *Stereotype Threat, Attributional Ambiguity, and Latino Underperformance.* Unpublished manuscript, University of Texas at Austin.

Bailey, T., Jeong, D.W., and Cho, S.W. (2010). Referral, enrollment, and completion in developmental education sequences in community colleges. *Economics of Education Review, 29*, 255–270.

Ballenger, C. (1997). Social identities, moral narratives, scientific argumentation: Science talk in a bilingual classroom. *Language and Education, 11*(1), 1–14.

Bang, M., Medin, D.L., and Atran, S. (2007). Cultural mosaics and mental models of nature. *Proceedings of the National Academy of Sciences of the United States of America, 104*(35), 13868–13873.

Bowen, W.G., Chingos, M.M., Lack, K.A. and Nygren, T.I. (2012). Interactive learning online at public universities: Evidence from randomized trials. *Ithaka S+R*, 1–52. Available: http://www.sr.ithaka.org/research-publications/interactive-learning-online-public-universities-evidence-randomized-trials [July 2015].

Brainard, S.G., and Carlin, L. (1998). A six-year longitudinal study of undergraduate women in engineering and science. *Journal of Engineering Education, 87*(4), 369–375.

Brandt, C.B. (2008). Discursive geographies in science: Space, identity, and scientific discourse among indigenous women in higher education. *Cultural Studies in Science. Education,* 3, 703–730.

Bressoud, D. (2014). *Attracting and Retaining Students to Complete Two- and Four-Year Undergraduate Degrees in STEM: The Role of Undergraduate Mathematics Education.* Commissioned paper prepared for the Committee on Barriers and Opportunities in Completing 2-Year and 4-Year STEM Degrees, National Academy of Sciences, Washington, DC. Available: http://sites.nationalacademies.org/cs/groups/dbassesite/documents/webpage/dbasse_088835.pdf [April 2015].

Bressoud, D., Carlson, M., Mesa, V., and Rasmussen, C. (2013). The calculus student: Insights from the MAA National Study. *International Journal of Mathematical Education in Science and Technology, 44*(5), 685–698. Available: http://www.researchgate.net/publication/260459087_The_calculus_student_Insights_from_the_MAA_national_study [April 2015].

Brown, B. (2004). Discursive identity: Assimilation into the culture of science and its implications for minority students. *Journal of Research in Science Teaching, 41*(8), 810–834.

Cabrera, A.F., Nora, A., Terenzini, P.T., Pascarella, E.T., and Hagedorn, L.S. (1999). Campus racial climate and the adjustment of students to college: A comparison between white students and African-American students. *Journal of Higher Education, 70,* 134–160.

Campuzano, L., Dynarski, M., Agodini, R., and Rall, K. (2009). *Effectiveness of Reading and Mathematics Software Products: Findings from Two Student Cohorts* (NCEE 2009-4041). Washington, DC: National Center for Educational Evaluation and Regional Assistance, Institute of Education Sciences, U.S. Department of Education.

Carleone H.B., and Johnson, A. (2007). Understanding the science experiences of successful women of color: Science identity as an analytic lens. *Journal of Research in Science Teaching, 44*(8), 1187–1218.

Carlson, M., Madison, B., and West, R. (2010). *The Calculus Concept Readiness (CCR) Instrument: Assessing Student Readiness for Calculus.* Available: arxiv.org/ftp/arxiv/papers/1010/1010.2719.pdf [July 2015].

Cech, E.A., and Waidzunas, T.J. (2011). Navigating the heteronormativity of engineering: The experiences of lesbian, gay, and bisexual students. *Engineering Studies, 1,* 1–24.

Chang, M.J., Eagan, M.K., Lin, M.H., and Hurtado, S. (2011). Considering the impact of racial stigmas and science identity: Persistence among biomedical and behavioral science aspirants. *Journal of Higher Education, 82*(5), 564–596.

Chemers, M.M., Zurbriggen, E.L., Syed, M., Goza, B.K., and Bearman, S. (2011). The role of efficacy and identity in science career commitment among underrepresented minority students. *Journal of Social Issues, 67*(3), 469–491.

Chen, X. (2009). *Students Who Study Science, Technology, Engineering, and Mathematics (STEM) in Postsecondary Education* (NCES 2009161). Washington, DC: U.S. Department of Education, Office of Educational Research and Improvement, National Center for Educational Statistics.

Cheryan, S., Plaut, V.C., Davies, P., and Steele, C.M. (2009). Ambivalent belonging: How stereotypical environments impact gender participation in computer science. *Journal of Personality and Social Psychology, 97,* 1045–1060.

Cobern, W.W., and Aikenhead, G. (1998). Cultural aspects of learning science. In B. Fraser and K. Tobin (Eds.), *International Handbook of Science Education* (Part One, pp. 39–52). Dordrecht, Netherlands: Kluwer Academic.

Cohen, G.L., Steele, C.M., and Ross, L.D. (1999). The mentor's dilemma: Providing critical feedback across the racial divide. *Personality and Social Psychology Bulletin, 25,* 1302–1318.

2003). An investigation of academic self-concept and its relationship to academic
ement in African American college students. *Journal of Black Psychology, 26*,
164.

, Nora, A., and Taggart, A. (2009). Student characteristics, pre-college, college,
environmental factors as predictors of majoring in and earning a STEM degree:
analysis of students attending a Hispanic-serving institution. *American Educational
search Journal, 46*, 924–942.

t, J., and Major, B. (1989). Social stigma and self-esteem: The self-protective properties
f stigma. *Psychological Review, 96*, 608–630.

., and Cromley, J.G. (2014). Changes in implicit theories of ability in biology and dropout from STEM majors: A latent growth curve approach. *Contemporary Educational
Psychology, 39*(3), 233–247.

ck, C.S. (2000). *Self-Theories: Their Role in Motivation, Personality, and Development.*
Philadelphia, PA: Taylor and Francis

veck, C.S., and Leggett, E.L. (1988). A social-cognitive approach to motivation and personality. *Psychological Review, 95*, 256–273.

ccles, J., Wigfield, A., and Schiefele, U. (1998). Motivation to succeed. In W. Damon (Series
Ed.) and N. Eisenberg (Volume Ed.), *Handbook of Child Psychology, Volume 3: Social,
Emotional, and Personality Development* (5th ed., pp. 1017–1095). New York: Wiley.

Espinosa, L.L. (2011). Pathways and pipelines: Women of color in undergraduate STEM
majors and the college experiences that contribute to persistence. *Harvard Educational
Review, 81*(2), 209–241.

Fleming, J. (1984). *Blacks in College: A Comparative Study of Students' Success in Black and
White Institutions.* San Francisco: Jossey-Bass.

Franklin, A.J., and Boyd-Franklin, N. (2000). Invisibility syndrome: A clinical model of the
effects of racism on African American males. *American Journal of Orthopsychiatry,
70*(2), 33–41.

Fries-Britt, S., and Griffin, K. (2007). The black box: How high achieving blacks resist stereotypes about black Americans. *Journal of College Student Development, 48*(5), 509–524.

Gainen, J. (1995). Barriers to success in quantitative gatekeeper courses. In J. Gainen and
E.W. Willemsen (Eds.), *New Directions for Teaching and Learning* (Issue 61, pp. 5–14).
San Francisco: Jossey-Bass

Gonzales P.M., Blanton H., and Williams K.J. (2002). The effects of stereotype threat and
double-minority status on the test performance of Latina/o women. *Personality and
Social Psychology Bulletin, 28*(5), 659–670.

Good, C. (2012). Why do women opt out? Sense of belonging and women's representation in
mathematics. *Journal of Personality and Social Psychology, 102*(4), 700–717.

Gurin, P., Dey, E.L., Hurtado, S., Gurin, G. (2002). Diversity and higher education: Theory
and impact on educational outcomes. *Harvard Educational Review, 71*(3), 332–366.

Hagerty, G., and Smith, S. (2005). Using the web-based interactive software ALEKS to enhance
college algebra. *Mathematics and Computer Education, 39*(3), 183–194.

Hagerty, G., Smith S., and Goodwin, D. (2010). Redesigning college algebra: Combining educational theory and web-based learning to improve student attitudes and performance.
Primus, 20(5), 418–437.

Herriott, S.R., and Dunbar, S.R. (2009). Who takes college algebra? *Primus, 19*(1), 74–87.

Hewlett, S.A., Luce, C.B., and Servon, L.J. (2008). *The Athena Factor: Reversing the Brain
Drain in Science, Engineering, and Technology.* Watertown, MA: Harvard Business
School.

Hope, E., Chavous, T.M., Jagers, R.J., and Sellers, R.M. (2013). Connecting self-esteem and
achievement: Diversity in academic identification and dis-identification patterns among
Black college students. *American Educational Research Journal, 50*(5), 1122–1151.

Hsu, E., Murphy, T.J., and Treisman, U. (2008). Supporting high achievement in introductory mathematics courses: What we have learned from 30 years of the Emerging Scholars Program. In M.P. Carlson and C. Rasmussen (Eds.), *Making the Connection: Research and Teaching in Undergraduate Mathematics Education.* MAA Notes #73. Washington, DC: Mathematical Association of America.

Hughes, R. (2012). Gender conception and the chilly road to female undergraduates' persistence in science and engineering fields. *Journal of Women and Minorities in Science and Engineering, 18*(3), 215–234.

Hurtado, S., Carter, D.F., and Spuler, A. (1996). Latino student transition to college: Assessing difficulties and factors in successful adjustment. *Research in Higher Education, 37*(2), 135–157.

Hurtado, S., Carter, D.F., and Kardia, D. (1998). The climate for diversity: Key issues for institutional self-study. *New Directions for Institutional Research, 25*(2), 53–63.

Hurtado, S., Eagan, M.K., Cabrera, N.L., Lin, M.H., Park, J. and Lopez, M. (2008). Training future scientists: Predicting first-year minority student participation in health science research. *Research of Higher Education, 49*(2), 126–152.

Hurtado, S., Eagan, M.K., Tran, M.C., Newman, C.B., Chang, M.J., and Velasco, P. (2011). We do science here: Underrepresented students' interactions with faculty in different college contexts. *The Journal of Social Issues, 67*(3), 553–579.

Hyde, J.S., and Mertz, J.E. (2009). Gender, culture, and mathematics performance. *Proceedings of the National Academy of Sciences of the United States of America, 106*(22), 8801–8807.

Johnson, D.R. (2011). Examining sense of belonging and campus racial diversity experiences among women of color in STEM Living-Learning Programs. *Journal of Women and Minorities in Science and Engineering, 17*(3), 209–223.

Johnson, D.R. (2012). Campus racial climate perceptions and overall sense of belonging among racially diverse women in STEM majors. *Journal of College Student Development, 53*(2), 336–346.

Kaufman, J., Ryan, C., Thille, C., and Bier, N. (2013). Open earning initiative courses in community colleges: Evidence on use and effectiveness. Menlo Park, CA: William and Flora Hewlett Foundation. Available http://www.hewlett.org/sites/default/files/CCOLI_Report_Final_1.pdf [July 2015].

Klingbeil, N.W., Mercer, R., Rattan, K.S., Raymer, M.L., and Reynolds, D.B. (2006). *Redefining Engineering Mathematics Education at Wright State University.* Paper presented at the 2006 Annual Conference and Exposition, American Society for Engineering Education, Chicago, IL. Available: https://peer.asee.org/redefining-engineering-mathematics-education-at-wright-state-university [April 2016].

Ko, L.T., Kachchaf, R.R., Hodari, A.K., and Ong, M. (2014). Agency of women of color in physics and astronomy: Strategies for persistence and success. *Journal of Women and Minorities in Science and Engineering, 20*(2), 171–195.

Koedinger, G., and Sueker, E. (1996). PAT goes to college: Evaluating a cognitive tutor for developmental mathematics. In *Proceedings of the Second International Conference on the Learning Sciences* (pp. 180–187). Charlottesville, VA: Association for the Advancement of Computing in Education.

Kurth, L.A., Anderson, C.A., and Palincsar, A.S. (2002). The case of Carla: Dilemmas of helping all students to understand science. *Science Education, 86*, 287–313.

Lee, L.A., Hansen, L.E., and Wilson, D.M. (2006). *The Impact of Affective and Relational Factors on Classroom Experience and Career Outlook among First-year Engineering Undergraduates.* Presentation at the Frontiers in Education 36th Annual Conference, Oct. 27–31, San Diego, CA.

Lemke, J.L. (2001). Articulating communities: Sociocultural perspectives on science education. *Journal of Research in Science Teaching, 38,* 296–316.

Leslie, S.J., Cimpian, A., Meyer, M., and Freeland, E. (2015). Expectations of brilliance underlie gender distribution across academic disciplines. *Science, 347*(6219), 262–265.

Litzler, E., Samuelson, C., and Lorah, J. (2014). Breaking it down: Engineering student STEM confidence at the intersection of race/ethnicity and gender. *Research in Higher Education, 55,* 810–832.

Locks, A.M., Hurtado, S., Bowman, N.A., and Oseguera, L. (2008). Extending notions of campus climate and diversity to students' transition to college. *Review of Higher Education, 31*(3), 257–285.

London, B., Rosenthal, L., Levy, S.R., and Lobel, M. (2014). The influences of perceived identity compatibility and social support on women in nontraditional fields during the college transition. *Basic and Applied Social Psychology, 33*(4), 37–41.

Malcom, L.E., Dowd, A.C., and Yu, T. (2010). *Tapping HSI-STEM Funds to Improve Latina and Latino Access to the STEM Professions.* Los Angeles: University of Southern California.

Malcom, S., Hall, P., and Brown, J. (1976). *The Double Bind: The Price of Being a Minority Woman in Science.* Washington, DC: American Association for the Advancement of Science.

Malcom, L.E., and Malcom, S.M. (2011). The double bind: The next generation. *Harvard Educational Review, 81*(2), 162–172.

McKay, P., Doverspike, D., Bowen-Hilton, D., and Martin, Q.D. (2002). Stereotype threat: Effects on the raven advanced progressive matrices scores of African Americans. *Journal of Applied Social Psychology, 32*(4), 767–787.

McGee, E.O., and Martin, D.B. (2011). "You would not believe what I have to go through to prove my intellectual value!" Stereotype management among academically successful black mathematics and engineering students. *American Educational Research Journal, 48*(6), 1377–1389.

Nadal, K., Wong, Y., Grifin, K.E., Davidoff, K., and Sriken, J. (2014). The adverse impact of racial microaggressions on college students' self-esteem. *Journal of College Student Development, 55*(5), 461–474.

National Academy of Engineering. (2012). *Colloquy on Minority Males in Science, Technology, Engineering, and Mathematics.* C. Didion, N.L. Fortenberry, and E. Cady (Rapporteurs). Washington, DC: The National Academies Press.

National Academy of Sciences, National Academy of Engineering, and Institute of Medicine. (2011). *Expanding Underrepresented Minority Participation: America's Science and Technology Talent at the Crossroads.* Committee on Underrepresented Groups and the Expansion of the Science and Engineering Workforce Pipeline. Committee on Science, Engineering, and Public Policy and Policy and Global Affairs. Washington, DC: The National Academies Press.

National Research Council. (2009). *Learning Science in Informal Environments: People, Places, and Pursuits.* Committee on Learning Science in Informal Environments, P. Bell, B. Lewenstein, A.W. Shouse, and M.A. Feder (Eds.). Board on Science Education, Center for Education, Division of Behavioral and Social Sciences and Education. Washington, DC: The National Academies Press.

National Research Council. (2011). *Expanding Underrepresented Minority Participation: America's Science and Technology Talent at the Crossroads.* Committee on Underrepresented Groups and Expansion of the Science and Engineering Workforce Pipeline, F.A. Hrabowski, P.H. Henderson, and E. Psalmonds (Eds.). Board on Higher Education and the Workforce, Division on Policy and Global Affairs. Washington, DC: The National Academies Press.

National Science Foundation and National Center for Science and Engineering Statistics. (2013). *Women, Minorities, and Persons with Disabilities in Science and Engineering: 2013*. Arlington, VA: National Science Foundation.

Nelson-Barber, S., and Estrin, E. (1995). *Culturally Responsive Mathematics and Science Education for Native Students*. Washington, DC: Nation Education Initiative of the Regional Educational Labs.

Nettles, M.T. (1988). *Toward Black Undergraduate Student Equality in American Higher Education*. New York: Greenwood Press.

Olitsky, S. (2006). Facilitating identity formation, group membership, and learning in science classrooms: What can be learned from out-of-field teaching in an urban school? *Science Education, 91*, 201–221.

Ong, M. (2002). *Against the Current: Women of Color Succeeding in Physics*. Doctoral dissertation, University of California, Berkeley, Pub. No. AAI3082349, Dissertation Abstracts International, Vol. 64-02, Section A. Available: http://adsabs.harvard.edu/abs/2002PhDT........27O [April 2016].

Ong, M. (2005). Body projects of young women of color in physics: Intersections of gender, race, and science. *Social Problems, 52*, 593–617.

Ong, M., Wright, C., Espinosa, L., and Orfield, G. (2011). Inside the double bind: A synthesis of empirical research on undergraduate and graduate women color in science, technology, engineering, and mathematics. *Harvard Educational Review, 81* (2), 172–208.

Osborne, J.W. (1997). Race and academic disidentification. *Journal of Educational Psychology, 89*(4), 728–735.

Osborne, J.W. (1999). Unraveling underachievement among African American boys from an identification with academics perspective. *Journal of Negro Education, 68*(4), 555.

Page, S.E. (2007). *The Difference: How the Power of Diversity Creates Better Groups, Firms, Schools, and Societies*. Princeton, NJ: Princeton University Press.

Palmer, R.T., Maramba, D.C., and Ii, T.E.D. (2011). A qualitative investigation of actors promoting the retention and persistence of students of color in STEM. *Journal of Negro Education, 80*(4), 491–504.

Pane, J.F., McCaffrey, D.F., Slaughter, M., Steele, J.L., and Ikemoto, G.S. (2010). An experiment to evaluate the efficacy of cognitive tutor geometry. *Journal of Research on Educational Effectiveness, 3*(3), 254–281.

Perez, T., Cromley, J.G., and Kaplan, A. (2014). The role of identity development, values, and costs in college STEM retention. *Journal of Educational Psychology, 106*(1), 315–329.

President's Council of Advisors on Science and Technology. (2012). *Report to the President. Engage to Excel: Producing One Million Additional College Graduates with Degrees in Science, Technology, Engineering and Mathematics*. Available: http://www.whitehouse.gov/sites/default/files/microsites/ostp/pcast-engage-to-excel-final_feb.pdf [April 2015].

Ramsey, L.R., Betz, D.E., and Sekaquaptewa, D. (2013). The effects of an academic environment intervention on science identification among women in STEM. *Social Psychology of Education, 16*(3), 377–397.

Rasmussen, C., and Ellis, J. (2013). Who is switching out of calculus and why? *Proceedings of the 37th Conference of the International Group for the Psychology of Mathematics Education*. Available: http://www.maa.org/sites/default/files/pdf/cspcc/rasmussen_ellis2013.pdf [October 2015].

Reid, L.D., and Radhakrishnan, P. (2003). How race still matters: The relation between race and general campus climate. *Cultural Diversity and Ethnic Minority Psychology, 9*, 263-275.

Reyes, M. (2011). Unique challenges for women of color in STEM transferring from community colleges to universities. *Harvard Educational Review, 81*(2), 241–263.

Riegle-Crumb, C., and King, B. (2010). Questioning a white male advantage in STEM: Examining disparities in college major. *Educational Researcher, 39,* 656–664.

Ritter, S. (2014). *Cognitive Tutor: Improvement toward STEM degrees.* Presentation to the Committee on Barriers and Opportunities in Completing 2-Year and 4-Year STEM Degrees, National Academy of Sciences, Washington, DC.

Sadler, P.M., and Tai, R.H. (2007). The two high-school pillars supporting college science. *Science, 317,* 457–458.

Scheines, R., Leinhardt, G., Smith, J., and Cho, K. (2005). Replacing lecture with web-based course materials. *Journal of Educational Computing Research, 32*(1), 1–26.

Schmader T., and Johns, M. (2003). Converging evidence that stereotype threat reduces working memory capacity. *Journal of Personality and Social Psychology, 85,* 440–452.

Sevilla, A., and Somers, K., (1993). Integrating precalculus review with the first course in calculus. *Primus, 3*(1), 35–41.

Seymour, E., and Hewitt, N. (1997). *Talking about Leaving: Why Undergraduates Leave the Sciences.* Boulder, CO: Westview Press.

Smith, D. (1997). *Diversity Works: The Emerging Picture of How Students Benefit.* Washington, DC: Association of American Colleges and Universities.

Smith, W.A., Hung, M., and Franklin, J.D. (2011). Racial battle fatigue and the "mis"education of Black men: Racial microaggressions, societal problems, and environmental stress. *Journal of Negro Education, 80*(1), 63–82.

Smith, J., Lewis, K., Hawthorne, L, and Hodges, S. (2013). When trying hard isn't natural: Women's belonging with and motivation for male-dominated STEM fields as a function of effort expenditure concerns. *Personality and Social Psychology Bulletin, 39*(2), 131–143.

Snipes, J., Fancsali, C., and Stoker, G. (2012). *Student Academic Mindset Interventions: A Review of the Current Landscape.* Washington, DC: IMPAQ International.

Solorzano, C.M., and Yosso, T. (2000). Critical race theory, racial microaggressions, and campus racial climate: The experiences of African American college students. *Journal of Negro Education, 69,* 60–73.

Sonnert, G., and Sadler, P.M. (2014). The impact of taking a college pre-calculus course on students' college calculus performance. *International Journal of Mathematical Education in Science and Technology, 45*(8), 1188–1207.

Steele, C.M. (1997). A threat in the air: How stereotypes shape intellectual identity and performance. *American Psychologist, 52,* 613–629.

Steele, C.M. (1992). Race and the schooling of African-American Americans. *The Atlantic Monthly, 269*(4), 68–78.

Steele, C.M. (1997). A threat in the air: How stereotypes shape intellectual identity and performance. *American Psychologist, 52,* 613–629.

Steele, C.M., and Aronson, J. (1995). Stereotype threat and the intellectual test performance of African Americans. *Journal of Personality and Social Psychology, 69,* 797–811.

Steele, C.M., Spencer, S.J., Hummel, M., Schoem, D., Carter, K., Harber, K., and Nisbett, R. (1998). Improving minority performance: An intervention in higher education. In C. Jencks and M. Phillips (Eds.), *Black-White Test Score Differences.* Cambridge, MA: Harvard University Press.

Strayhorn, T.L. (2010a). When race and gender collide: Social and cultural capital's influence on the academic achievement of African American and Latino males. *The Review of Higher Education, 33*(3), 307–322.

Strayhorn, T.L. (2012). *College Students' Sense of Belonging: A Key to Educational Success for All Students.* New York: Routledge.

Thompson, P.W., Castillo-Chavez, C., Culbertson, R., Flores, A., Greeley, R., Haag, S., et al. (2007). *Failing the Future: Problems of Persistence and Retention in Science, Technology, Engineering, and Mathematics Majors at Arizona State University.* Provost Office Report. Tempe: Arizona State University.

Tyson, W. (2011). Modeling engineering degree attainment using high school and college physics and calculus coursetaking and achievement. *Journal of Engineering Education. 100*(4), 760–777.

Williams, M.M., and George-Jackson, C.E. (2014). Using and doing science: Gender, self-efficacy, and science identity of undergraduate students in STEM. *Journal of Women and Minorities in Science and Engineering, 20*(2), 99–126.

Wilson, T.D., and Linville, P.W. (1985). Improving the performance of college freshmen with attributional techniques. *Journal of Personality and Social Psychology, 49*, 287–293.

Wolcott, B. (2001). Role models needed. *Mechanical Engineering, 123*(4), 47–51.

Yosso, T.J. (2000). *A Critical Race and LatCrit Approach to Media Literacy: Chicana/o Resistance to Visual Microaggressions.* Unpublished doctoral dissertation, University of California, Los Angeles.

Yosso, T.J. (2005). Whose culture has capital?: A critical race theory discussion of community cultural wealth. *Race, Ethnicity and Education, 8*(1), 69–91.

Yosso, T.J., Smith, W.A., Ceja, M., and Solorzano, D.G. (2009). Critical race theory, racial microaggressions, and campus racial climate for Latina/o undergraduates. *Harvard Review, 79*(4), 659–691.

4

Instructional Practices, Departmental Leadership, and Co-Curricular Supports

---

Major Messages

- Adoption of reformed curriculum and reformed teaching practices remains difficult because of such barriers as little support from other faculty and the department, few incentives for improved teaching, inappropriate classroom infrastructure, limited awareness of research-based instructional practices, and lack of time. Departments are a critical unit for change in undergraduate science, technology, engineering, and mathematics (STEM) education since they represent not only individual faculty values and aspirations, but also the curriculum as a whole beyond the individual courses that faculty teach.

- Co-curricular supports, if done well, can provide authentic disciplinary experiences and attend to the social and relational aspects of learning that have been shown to influence students' academic engagement and persistence.

Research conducted across all disciplines, not just STEM, indicates that the faculty behaviors and characteristics that have a significant effect on student engagement include active and collaborative learning techniques, communicating high expectations to students, course-related student-faculty interactions, and an emphasis on enriching educational experiences (Umbach and Wawrzynski, 2005). Thus, the educational context created by faculty behaviors and attitudes affect student learning and engagement. Two key features of the educational context are the instructional strategies and classroom environments that students encounter. Addressing curriculum and classroom concerns is a necessary component in any undergraduate STEM education effort. In this chapter, we focus on the barriers and opportunities to improving STEM teaching practices. In doing so, we describe the role that faculty, departments, and institutions can play in instructional reform. We also point to a set of strategies, beyond curricular reform, that can support persistence and completion of STEM credentials.

Throughout the chapter, we stress that instructional reform is not sufficient in and of itself. The learning environment, the culture of a department, the need for community, and the other factors described in Chapter 3 also play crucial roles. For example, Ko and colleagues (2014) have found that the messages that women of color often receive—directly or indirectly—from their academic settings (e.g., interactions with faculty, advisor, and peers; structure of departments; and classroom norms) convey low expectations, stereotypical views, and benign racism/sexism. Additionally, as is discussed later in the report, the policies that shape actions by faculty, departments, and institutions are also critical elements in creating an environment that can support success in STEM for all students by addressing cultural, instructional, and institutional policy barriers.

IMPROVING STEM TEACHING PRACTICES

Instructional strategies in undergraduate STEM classrooms matter. The most comprehensive meta-analysis to date illustrates that students learn more in STEM classrooms where instructors use active learning strategies rather than traditional lecturing (Freeman et al., 2014). A review of discipline-based education by the National Research Council (2012) revealed similar findings: that traditional lectures are less effective then evidence-based instructional strategies at improving conceptual knowledge and attitudes about learning STEM. The report illustrated that evidence-based instructional strategies include a range of approaches, including making lectures more interactive, having students work in groups, providing formative feedback, and incorporating authentic problems and activities. In particular, the report emphasizes that instructors' clarifying and facilitating student conceptual understanding is relevant across all STEM fields. While approaches to problem solving differ across fields, most research indicates that authentic problems and appropriately sequenced experiences are important for student learning of core concepts in STEM (National Research Council, 2012).

The National Research Council's report (2012) also found that active instructional strategies supported all students' STEM learning, and they especially supported learning among underrepresented students. Research on an active-learning intervention in physics and biology illustrates the disproportionally positive effect of a moderately structured intervention on black and first-generation college students (Beichner et al., 2007; Eddy and Hogan, 2014): the achievement gap between black and white students was halved, and the achievement gap between first-generation and other students was eliminated.

More nuanced studies are now being funded to identify the elements of successful instruction and how the elements may differ across groups (Eddy and Hogan, 2014). Even with more nuanced evidence, evidence-based approaches to teaching may be difficult to implement (National Research Council, 2012; Freeman et al., 2014; Eddy and Hogan, 2014). The complexity of the demands of faculty work in the 21st century, regardless of institution type, creates challenges to changing approaches to teaching.

To understand how teaching approaches are developed and codified, it is important to understand that teaching practices are situated in the context of departments and disciplinary norms, perceptions of how students learn, faculty values, pedagogical strategies, and faculty views of the impact of their teaching choices (Austin, 2011). Cultivating change in teaching practice is not as simple as demonstrating research evidence of instructional effectiveness: it also has to be linked to faculty experience, appointment type, disciplinary understanding, and departmental culture (Austin, 2011). In this section, we focus specifically on the nature of research-based STEM instructional strategies and the barriers and opportunities to implementing and sustaining this kind of instruction.

Significant resources have been invested in disseminating "best practices" in instruction (for an overview, see National Research Council, 2012, 2013). Disciplinary societies have made resources for improving teaching available to faculty through online archives or warehouses such as COMPADRE in physics (Mason, 2007) and the Advance Technology Education Program's National Resource Center;[1] an increasing number of disciplinary-based journals offer peer-reviewed research about effective practices; and a number of professional organizations make available professional development opportunities for faculty to learn about and practice new pedagogies (see Hilborn, 2013). The field of chemistry has been particularly successful in the application of socially mediated teaching and learning as evidenced by Process Oriented Guided Inquiry Learning (POGIL) and Peer Led Team Learning (PLTL), both of which use small groups of peer-led teams in problem solving.[2] There are barriers associated with these dissemination efforts, but they offer a clear opportunity for

---

[1]For more information, see https://atecentral.net/resources [July 2015].
[2]For more information, see http://www.pogil.org and http://www.pltl.org [April 2015].

faculty to learn about and adopt research-based instruction in most STEM disciplines. See Appendix A for an overview of some current instructional reform efforts in STEM fields.

BARRIERS TO INSTRUCTIONAL CHANGE
FACED BY STEM FACULTY

Teaching, research, and service represent the traditional three-legged stool that defines faculty work. The specific context within which this work is carried out is related to faculty decision making and practice relative to their allocation of time and effort. Institutional context, departmental structure and leadership, institutional incentives, and professional development opportunities determine faculty motivation to consider evidence-based approaches to teaching and student learning rather than their own experiences and department tradition.

According to a survey conducted during the 2013–2014 academic year, faculty, including faculty in STEM departments, have increased their use of evidence-based instructional strategies (Eagan et al., 2014). Full-time faculty reported that over the past 25 years, they increased their use of classroom discussions (from 70% to just over 80%), of group projects (from under 20% to 45%), of cooperative learning (from about 25% to 61%), and of student evaluation of each other's work (from about 10% to over 40%). However, 51 percent of full-time faculty continue extensive use of lecturing. There are a handful of studies of the instructional strategies in two STEM disciplines: physics and engineering. These studies indicate that widespread changes have not been adopted (Borrego et al., 2010; Henderson, 2008; Henderson and Dancy, 2009; Henderson et al., 2012; Prince et al., 2013). For example, a survey of physics faculty revealed that one-third of physics faculty do not use any evidence-based instructional strategies, one-third use one or two strategies, and one-third use at least three strategies (Henderson et al., 2012). A survey of engineering departments indicates that awareness of evidence-based teaching strategies is much higher than adoption (82% and 47% respectively) (Borrego et al., 2010). The results of these studies should be interpreted with caution, because faculty have been found to over-report their use of evidence-based instructional strategies, and there may be selection bias in which faculty members respond (Dancy and Henderson, 2010; Savkar and Lokere, 2010).

The rate of change in instructional strategies can be understood in terms of a set of barriers faced by the academic STEM community. The most general set of barriers is related to the lack of institutional incentives that faculty members have to adopting research-based instructional strategies or more innovative curricular programs. Such barriers as research time versus teaching time, faculty workloads, and resources can affect faculty decisions to invest in new teaching practices (Fairweather, 2008).

The preparation and professional development related to instructional strategies that STEM instructors have received can also be a barrier to implementing evidence-based strategies. Faculty members bring to their work a socialization that occurs during graduate education, particularly with respect to their identities as teachers and scholars (Austin, 2010). Centers for teaching and learning have been developed to provide collaborative networks across institutions. For example, the Center for Integration of Research, Teaching, and Learning (CIRTL), which is funded by the National Science Foundation (NSF), emphasizes preparing STEM future faculty to bring their scholarship to teaching and develop learning communities for professional development at both the institutional and national levels. CIRTL has also recognized the importance of learning skills that leverage the increasing student diversity in STEM classrooms and research environments as a mechanism to enhance educational excellence.[3]

Often, the approaches used to encourage faculty to adopt research-based curricula have not been effective. In the "develop and disseminate" model of change identified by Dancy and Henderson (2010), faculty members are expected to consider adopting a research-based curriculum on the basis of attending a 1-day workshop or other relatively short-time dissemination efforts. The National Science Foundation and other granting agencies previously supported this approach by often requiring the grantees to run workshops on developed curricula or carry out other forms of dissemination (Seymour, 2001). Although a very large number of STEM faculty members may have attended a dissemination workshop, it has not correlated with a large move toward adoption of STEM educational reforms (Borrego and Henderson, 2014; Henderson, 2008; Henderson et al., 2011). The National Science Foundation has moved away from the "develop and disseminate" approach in its recent program solicitations (e.g., Transforming Undergraduate Science Education, and Course Curriculum and Laboratory Improvement).

More successful approaches to training faculty, such as summer institutes and new faculty workshop series are now being implemented. One of the longest running new faculty professional development workshops is Project NExT (New Experiences in Teaching),[4] run by the Mathematical Association of America. Since 1994, it has served more than 1,500 new mathematics faculty. The 2-year program provides new faculty with a series of teaching workshops and a network of peer mentors. Another program

---

[3] For more information, see http://www.cirtl.net/ [July 2015].

[4] For more information, see http://www.maa.org/programs/faculty-and-departments/project-next [July 2015].

is run by the American Association of Physics Teachers (AAPT),[5] which has workshops for physics, astronomy, and engineering faculty (Felder and Brent, 2010) that provide new faculty with the opportunity to exchange experiences and tools.

In general, effective faculty development workshops incorporate content drawn from discipline-specific education research, involve discipline-specific educators as facilitators or co-facilitators, and address a need for sustainable support (Felder et al., 2011). For example, at AAPT's new faculty workshop, a small number of techniques that have proven to be effective in a variety of environments are presented. The workshops are meant to focus on tactics that can be implemented with minimal time and effort, thus allowing new faculty to better balance their teaching, research, and scholarship. In 2014, the workshops covered such topics as interactive lectures, peer instruction, just-in-time teaching, research in physics education, problem solving, and teaching for retention and diversity. The Howard Hughes Medical Institute and the National Academies of Sciences, Engineering, and Medicine also have partnered to run summer institutes to develop the teaching skills of faculty and instructional staff.[6]

The NSF's Advanced Technological Education (ATE) Program has generated a wide range of professional development resources for instructors involved in technician education, including problem-based learning, linkages with industry, career exploration and advising, and instructing diverse student groups. One ATE-supported effort, TeachTechnicians.org, increases access to and participation in faculty professional development: it is designed to be a one-stop shop for professional development opportunities provided by ATE grantees and others. The site provides ATE grantees a central place to announce and promote professional development events. It also provides grantees with access to expertise, vetted resources, and successful practices that they can use to improve technician education at their institution.

Even among those who have adopted new approaches, sustainability can be an issue. A study of research-based instructional strategies in introductory physics classrooms during fall 2008 found that long-term adoption of such strategies is hampered by discontinued funding for curriculum reform efforts and insufficient support from colleagues during implementations (Henderson et al., 2012). Research is needed to assess whether these factors are also barriers to adoption in other STEM fields.

Once a faculty member has decided to implement research-based instruction, she or he faces multiple barriers to implementing the instruction

---

[5]For more information, see http://www.aapt.org/conferences/newfaculty/nfw.cfm [May 2015].

[6]For more information, see http://www.academiessummerinstitute.org/ [July 2015].

with fidelity. Beyond awareness of and familiarity with the instructional strategy, an individual faculty member is often not fully aware of all the elements required for successful implementation. These might include skills in guiding student discourse (Duschl, 2002), engaging in the appropriate form of dialogue with the students, and avoiding microaggressions and implicit bias (Cohen et al., 1999; Hurtado et al., 2011; Nadal et al., 2014). Faculty members may also face situational-based barriers (Henderson, 2008), including not being able to cover as much content as when lecturing, possibly needing more tutorial sections, and scheduling constraints due to the need for particular classrooms that support collaborative work (see Box 4-1

BOX 4-1
Classroom Infrastructure

Class size and physical space can influence the extent to which faculty members apply evidence-based teaching strategies. Research in physics (Henderson and Dancy, 2007) and geoscience education (Macdonald et al., 2005) shows that large class sizes and the traditional classroom space may act as barriers to the adoption of innovative teaching approaches by faculty. Some classroom reforms call for major changes in room size and structure, for example (National Research Council, 2012, p. 127):

[T]he Student-Centered Active Learning Environment for Undergraduate Program (SCALE-UP) begins with a redesign of the classroom. Each room holds approximately 100 students, with round tables that accommodate 3 laptops and 9 students, whiteboards on several walls, and multiple computer projectors and screens so every student has a view. Students engage in hands-on activities and with computer simulations, work collaboratively on problems, and conduct hypothesis-driven experiments. SCALE-UP students have better scores on problem-solving exams and concept tests, slightly better attitudes about science, and less attrition than students in traditional courses (Beichner et al., 2007; Gaffney et al., 2008).

Another well-known reform is Studio Physics (National Research Council, 2012, p. 127):

Studio Physics redesigned teaching spaces to accommodate an integrated lecture/laboratory course. Early studies showed little improvement in students' conceptual understanding or problem-solving skills, despite the popularity of the innovation. Later implementations, which added research-based curricula, resulted in improved learning of content over traditional courses (Cummings et al., 1999; Sorensen et al., 2006), but not always improvements in problem solving (Hoellwarth et al., 2005).

for an overview of research on classroom design). In addition, a study of calculus instruction (Bressoud et al., 2013) indicates that it may be more difficult for faculty who do not employ good general instructional practices to shift to active instructional strategies because students sometimes are unhappy with and resist such strategies. Finally, faculty members who choose to make significant curricular changes without a support network of local colleagues and their departments are at an immediate disadvantage (Beach et al., 2012).

AUTHENTIC STEM EXPERIENCES

The President's Council of Advisors on Science and Technology (2012) and others (see, for example, Kuh, 2008) have stressed that exposure to authentic STEM experiences, including research, is a key aspect in improving persistence and completion. Authentic undergraduate STEM experiences can involve hypothesis-driven, hands-on experimentation in which the outcome is unknown, peer-to-peer support, faculty-student interactions, and academic support. Students can be exposed to authentic STEM experiences in myriad ways, but typically students are provided such experiences via course-based opportunities to do investigations or by participating in a faculty's research laboratory.

Classroom-based strategies that engage students in authentic STEM experiences are in line with evidence-based instructional strategies that require moving away from lectures and recipe-based laboratory exercises toward more open-ended and student-driven STEM experiences (National Research Council, 2012). Evidence exists on the value of integrating authentic STEM experiences via undergraduate research and project-based laboratories (National Research Council, 2012; President's Council of Advisors on Science and Technology, 2012; Weaver et al., 2008). Such activities can be included in the curriculum of the undergraduate STEM laboratory or structured research programs, such as the Minority Biomedical Research Support (MBRS) Program or Maximizing Access to Research Careers (MARC), which are supported by the National Institutes of Health (Eagan et al., 2013).

Undergraduate research programs and internships may be particularly important for students from underrepresented groups since they may facilitate students' identities as scientists and engineers (Eagan et al., 2013). Authentic experiences may also involve opportunities to work on industry-related projects, as in the successful engineering clinic program at Harvey Mudd College.[7] Begun in 1963, the program has become an integral part of the college's engineering program and involves undergraduates at all levels.

---

[7]For more information, see https://www.hmc.edu/clinic/ [July 2015].

It engages small groups of undergraduate students working on industry-sponsored design projects.

In 2015, the National Academies of Sciences, Engineering, and Medicine organized a convocation to explore many aspects of the opportunities and challenges of introducing various models of discovery-based approaches to STEM education into undergraduate curricula.[8] Another committee is currently conducting a consensus study on these issues, with a report expected in 2016.

TENURE-TRACK AND CONTINGENT STEM FACULTY APPOINTMENTS

The nature of faculty appointments is also a factor in the learning environment that STEM students encounter. Both NSF and the U.S. Department of Education collect data on undergraduate faculty including faculty in STEM departments, but information on nontenure-track faculty and staff has not been available since the Department of Education discontinued the National Study of Postsecondary Faculty in 2004. However, studies of undergraduate STEM instructors and surveys of instruction conducted by disciplinary societies provide a partial picture of the contributions of tenured, tenure-track, and nontenure-track faculty and staff to student's learning.

The continuing change in balance from permanent tenure-track appointments that include all aspects of faculty work—teaching, research, and service—to nontenure-track, fixed-term, contingent, and part-time positions that emphasize only instruction may in effect marginalize the significance of teaching (Austin, 2011). This shift may convey the idea that teaching is less important than the other aspects of faculty work and disconnect teaching from research and the culture and community of the field.

Instructors with different types of appointments are teaching major parts of the undergraduate curriculum across all disciplines, even at the important introductory level (Baldwin and Wawrzynski, 2011). Teaching practices of part-time contingent faculty differ from those of other faculty. In a study of faculty at 4-year institutions from all academic departments, part-time faculty interacted with students less often, used active and collaborative instructional strategies less frequently, had lower academic expectations, and spent less time preparing for classes than did full-time faculty (both tenure-track and nontenure-track) (Baldwin and Wawrzynski, 2011).

Within STEM disciplines, it has been argued that part-time faculty in introductory gatekeeper courses can affect students' engagement and persistence. Some believe that students have fewer meaningful interactions with

---

[8]For more information, see http://dels.nas.edu/Past-Events/Convocation-Integrating-Discovery-Based-Research/AUTO-9-90-18-T [July 2015].

part-time faculty, which leads students to be less integrated into academic culture and thus be negatively affected in terms of persistence. Part-time faculty members are typically limited in their ability to engage students in research experiences, because of time constraints and because they do not conduct research at the college or university. One study by Eagan and Jaeger (2008) found that students were significantly and negatively affected by having gatekeeper courses taught by part-time faculty.

Community college students enrolled in STEM courses have a high probability of taking courses taught by part-time faculty, and instruction by part-time faculty is negatively correlated with student retention and transfer to a 4-year institution (Jaeger and Eagan, 2009, 2011). Students with greater levels of exposure to part-time faculty are less likely to earn an associate's degree in comparison with students who do not receive any instruction by part-time faculty (Jaeger and Eagan, 2009). Particularly in the sciences, a first-year student who has spent more than the average amount of time with part-time instructors is less likely to transfer to a 4-year institution than a classmate who has not had a part-time instructor (Jaeger and Eagan, 2011).

The American Chemical Society (ACS) Committee on Professional Training surveyed chemistry programs at 4-year institutions in 2010 in order to understand the effects of nontenure-track appointments on undergraduate chemistry education (American Chemical Society, 2010).[9] The results indicated that 66 percent of general chemistry lecture courses for majors were taught by tenure-track faculty, while just 30 percent of general chemistry lecture courses for nonmajors were taught by tenure-track faculty. A similar trend was found in organic chemistry classes; tenure-track faculty taught 80 percent of courses for majors and they taught 50 percent of courses for nonmajors. In addition, the ACS 2010 report indicates that laboratory instruction was primarily done by contingent chemistry faculty. Trends such as these suggest that primary instruction by nontenure-track faculty who do not have access to ongoing research programs may present a barrier to students interested in furthering their research experience.

DEPARTMENTAL LEADERSHIP AND STEM INSTRUCTION

Department leadership has the capacity to enhance instructional strategies and support for STEM student learning. The department is the critical unit for change in undergraduate STEM education since it represents not only individual faculty values and aspirations, but also the curriculum as an integral whole beyond individual courses. Departmental commitment is critical for the continuous assessment of teaching practices and support for

---

[9] The survey specifically excluded teaching assistants.

experimentation and innovation. Individual faculty investment in new pedagogical approaches cannot be sustained or spread by itself, and institution-wide programs are often too diluted. The department is the practical unit that can affect change because it has the authority to establish on-campus programs that explicitly recognize high-quality instruction.

There are many "levers" that department leaders can use to drive change, including setting learning goals, adjusting prerequisites, increasing flexibility of class taking, providing incentives and rewards for improved pedagogy, revising teaching assignments, providing support for course re-design, and reviewing when classes are offered. STEM departments can create teaching awards, offer access to the resources and release time needed by faculty to engage in educational endeavors, and provide recognition of those endeavors in promotion and tenure decisions (Brewer and Smith, 2010).

Departmental efforts to create change can be hampered by the lack of data available to inform reform decisions. Without reliable information about where students encounter barriers, the nature of the barriers, and profiles of the students who encounter barriers, it can be difficult for leaders to determine what actions to take. Some universities have begun to address the need for reliable data by partnering with the institutional offices and divisions that have access to student data (i.e., institutional research centers) and by developing easy-to-use data analysis and visualization tools.[10]

Physics has provided an interesting platform to examine the effectiveness for the department as the unit for change in STEM undergraduate education. A national task force on undergraduate physics through the American Institute of Physics, the American Physical Society, and the American Association of Physics Teachers examined the characteristics of "thriving" departments (Hilborn and Howes, 2003). The common elements across departments included a well-developed curriculum, individualized advising and mentoring, an undergraduate research program or industry-based internships (or both), many opportunities for informal student-faculty interactions, and a strong sense of community supported by departmental leadership across faculty and students. For details on efforts to create and sustain change in undergraduate life science education, see Box 4-2.

In 2007, the American Association for the Advancement of Science hosted a series of regional meetings to discuss what needed to be done to improve undergraduate biology education. The meetings were attended by over 200 biology faculty, college and university administrators, and other undergraduate biology stakeholders. The input from these meetings was used to frame a 2009 national conference on undergraduate biology reform.

---

[10] For an example of an award-winning program, see http://iamstem.ucdavis.edu/tools/ [July 2015].

BOX 4-2
The Partnership for Undergraduate
Life Science Education Project

The Partnership for Undergraduate Life Science Education (PULSE) Project grew out of the report on undergraduate biology education *Vision and Change* (American Association for the Advancement of Science, 2011). It has focused on an inclusive, student-centered, evidence-based teaching and learning approach. It has identified the department as the critical unit for change. Through the work of the PULSE community, a framework for examining departmental change for core issues, such as student metacognitive skills, authentic research experiences, pedagogical approaches, faculty development, and assessment and the resources and tools for initiating change in these areas have been identified.*

---

*For details, see http://www.pulsecommunity.org/ [April 2015].

The conference was attended by over 500 biology faculty, college and university administrators, and other undergraduate biology stakeholders. The conference focused on six major questions: (1) what undergraduates in biology should know and be able to do, (2) how should students be taught, (3) how should learning be assessed, (4) how should professional development of instructors be conducted, (5) what institutional changes are needed, and (6) what tools are needed to facilitate change. The conference yielded the following action steps that biology departments across the country are working to implement (Brewer and Smith, 2010, p. 50):

- Mobilize all stakeholders, from students to administrators, to commit to improving the quality of undergraduate biology education.
- Support the development of a true community of scholars dedicated to advancing the life sciences and the science of teaching.
- Advocate for increased status, recognition, and rewards for innovation in teaching, student success, and other educational outcomes.
- Require graduate students who are on training grants in the biological sciences to participate in training in how to teach biology.
- Provide teaching support and training for all faculty, but especially postdoctoral fellows and early-career faculty, who are in their formative years as teachers.

CO-CURRICULAR STRATEGIES FOR
IMPROVING STEM EDUCATION

As outlined by Estrada (2014), co-curricular supports,[11] if done well, provide authentic disciplinary experiences while also taking into account the social and relational aspects of learning that have been shown to influence students' academic engagement and persistence in the sciences (Chang et al., 2011; Kinkead, 2003; Lopatto, 2003). Specifically, co-curricular programming can mitigate the negative psychological and academic impacts of a stigmatizing STEM academic culture by affirming students' self-perceptions of competence (Gandara and Maxwell-Jolly, 1999; Hurtado et al., 2009; Mabrouk and Peters, 2000) and sense of community in the college setting. Thus, such programming can serve important roles both in promoting motivation and achievement and in protecting students when they experience stigma and exclusion.

STEM faculty members and leaders of co-curricular reforms have to be supported by their departments and institutions through allocation of time, resources, and other types of support. Once STEM reform begins, the need for support continues as the co-curricular reform requires subsequent adaptations and modifications. Payoff in the form of improved learning outcomes may not be apparent in early stages of such efforts but should be expected later. That is, administrators and faculty need to be aware of and accept that a significant proportion of the costs of innovation will be at the beginning and that a sustained effort will be required to support the reform effort over multiple years. Everyone involved in reform efforts needs to have realistic temporal and financial expectations for anticipated outcomes. This section provides a basic overview of key elements in that reform; we provide a detailed discussion of creating and sustaining systemic change in Chapter 6.

Internships

As discussed above, internships provide important opportunities for students to have hands-on experiences in their fields. Internships provide an opportunity to expand on the learning community developed in a student's program through sustained engagement with people working in industry (Eagan, 2013). There is some evidence that participation in an internship is significantly correlated to persistence in undergraduate engineering and computer science (Eagan, 2013). According to Fifolt and Searby (2010, p. 21), "mentoring students and new graduates can provide a bridge be-

---

[11] Co-curricular supports are activities, programs, and learning experiences that complement, in some way, what students are learning in the classroom.

tween theory learned in college and the complex realities of the workforce environment." When structured properly, internships provide students with this valuable mentorship experience. Internships can be research or design based or focused on working in an organization, catering to the wide array of opportunities that are available to STEM majors and providing students with the option to explore different career paths. Regardless of the type, well-run internships expose students to authentic research or design activities and hands-on experiences through "a mutual process of discovery that occurs through dialogue and activity" (Thiry et al., 2011, p. 361).

Some colleges and universities actively promote such opportunities through partnerships with local companies. For example, when officials at Miami-Dade College proposed a new B.S. degree in information systems technology, they secured an agreement from Florida Power and Light to provide internships to undergraduates in the program.[12] Florida Power and Light also provides internships to some students studying for associate's degrees in electrical power technology.[13]

Summer Bridge Programs

Summer bridge programs can enhance the precollege experience of all students, helping them become familiar with STEM-related curricula, academic expectations, program structure, peers and faculty, and career opportunities. Summer bridge programs have been demonstrated to have a positive effect on retention, especially among students from traditionally underrepresented groups (Strayhorn, 2010b). Summer bridge programs that cater to STEM disciplines have been shown to enhance student success (Association of American Colleges and Universities, 2012; Gilmer, 2007). To best prepare students to succeed in STEM disciplines, STEM-related summer bridge programs should take a multipronged approach, including a combination of activities and programming that address their "academic, social, and career needs" (Lenaburg et al., 2012, p. 153). Specific elements include an orientation to campus life and resources, an introduction to research activity and presentation, mentoring programs that connect new and prospective students with current students, and a structured session that engages students in career exploration (Lenaburg et al., 2012). Programs that integrate these different elements will provide students not only with a sense of community, but also with the tools necessary to succeed in college.

For example, at North Carolina State University, the Women in Science

---

[12] For details, see http://www.nexteraenergy.com/employeecentral/emp_comm/docs/ENG0509.pdf [April 2015].

[13] For details, see https://www.mdc.edu/homestead/pdf/EPT_Program%20Sheet.pdf [April 2015].

and Engineering (WISE) Program[14] offers students the chance to move into their dorm rooms a few days early to participate in a summer bridge program. The goal of the program is to provide support for the students that will ease their transition to college. The students participate in group work where they do hands-on activities to stimulate the use of problem-solving skills and creativity. They are assigned an upper-class mentor and engage in discussions about academics and campus life.

Participants in the WISE Summer Bridge are entering members of the WISE Village, a living and learning community designed especially for first- and second-year women majoring in science or engineering at North Carolina State. The WISE Village plans social, educational, and cultural activities to help residents interact with each other and develop a sense of community while exploring some of the opportunities available to them at North Carolina State. In another component of the program, WISE offers free tutoring in calculus, chemistry, and physics, three nights a week in the common dorm, where the students study together with the assistance of their mentors and tutors. An assessment of the WISE Program shows that participants are retained in the sciences and engineering at a higher rate than their non-WISE counterparts (Titus-Becker et al., 2007). Graduation rates of WISE participants could not yet be calculated because 4-year graduate rates on the first cohort were not yet available (the program was in its fifth year at the time the assessment took place).

Student Professional Groups

Many disciplinary professional societies and societies for professionals from underrepresented groups now include student chapters. The student chapters are oriented toward building community among members, connecting members to STEM professionals, and developing members' disciplinary identity. For example, the National Society for Black Engineers has a collegiate membership[15] category that allows student members access to networking, conferences, career fairs, test-preparation workshops, tutoring, and scholarship opportunities. A collegiate membership in the Society of Women Engineers includes access to career guidance, networking events, leadership trainings, and professional development seminars.[16] In addition, the Society for the Advancement of Hispanics/Chicanos and Native Americans in Science (SACNAS) offers student memberships that link students

---

[14]For details, see http://www.ncsu.edu/wise/bridge.htm [April 2015].

[15]For details, see http://www.nsbe.org/Membership/Membership-Benefits.aspx#.VOY5R_nF-VM [April 2015].

[16]For details, see http://societyofwomenengineers.swe.org/membership/benefits-a-discounts/409-membership-types/3361-collegiate-membership [May 2015].

to a national network of mentors and peers, provides access to electronic magazines and newsletters of the society, and allows participation at a national conference.[17] In addition, some campuses have local student chapters of national organizations, many with active Facebook groups promoting campus meetings and activities.[18]

Peer Tutoring

Peer tutoring involves people in similar social groupings, who are not professional teachers, working together to learn. Traditionally, peer tutoring has been thought of as a knowledgeable student transmitting knowledge to a less knowledgeable student. A wide range of peer-tutoring formats has developed over the past decade. Peer-tutoring formats vary across a number of dimensions, including the ratio of tutors to tutees, ability or knowledge of the tutor and tutee, and the amount of tutoring time (Topping, 1996). There is substantial evidence on the effectiveness of the various formats of peer tutoring (Topping, 1996), for both the tutor and the tutee (Annis, 1983; Benware and Deci, 1984) in terms of academic achievement (American River College, 1993; Lidren et al., 1991), self-efficacy (Schunk, 1987), and motivation (Schunk, 1987).

For example, California State University, San Marcos (CSUSM) operates a 35 hour-a-week drop-in STEM tutoring center.[19] Undergraduate tutors in math and science support students enrolled in lower-division gateway STEM courses. The STEM tutoring program at CSUSM benefits both tutors and students. For students seeking assistance, the tutors provide timely course-related assistance. Tutors also encourage students to work together, fostering a sense of community. This can help students establish peer networks that persist beyond the tutoring center and may form the basis of informal student learning communities (Cooper, 2010). For tutors, tutoring provides flexible employment for high-achieving upper-division science and mathematics majors. In addition to deepening their own content knowledge, tutors develop communication skills and gain an appreciation for teaching and learning that is applicable to graduate school and future careers (Arco-Tirado et. al., 2011; Topping, 1996).

---

[17] For details, see http://sacnas.org/community/membership/benefits [May 2015].

[18] For examples, see http://societyofwomenengineers.swe.org/membership/benefits-a-discounts #activePanels_0 [May 2015] and http://www.acs.org/content/acs/en/education/students/college/ studentaffiliates.html, and https://awis.site-ym.com/?ChapterDuesList [May 2015].

[19] For details, see http://www.csusm.edu/stem/stemcenter.html [April 2015].

Living and Learning Environments

To address the connection between successful transition to college and students' engagement with and connection to their college community (see Astin, 1984; Pascarella and Terenzini, 2005), an increasing number of institutions have created living-learning programs. Living-learning programs cluster students with shared academic goals or focus in residential communities (Shapiro and Levine, 1990). Four major types of learning communities have been identified: paired or clustered courses; cohorts in large courses or first-year interest groups; team-taught courses; and residential learning communities (Inkelas et al., 2008; Shapiro and Levine, 1990).

Living-learning programs at several campuses have been correlated to positive transition to college and positive academic outcomes (Pike, 1999; Pike et al., 1997; Stassen, 2003). The strength of the evidence varies by the type of living-learning communities, type of institution, discipline, and student characteristics. Successful living-learning programs tend to share three characteristics: a strong presence and partnership with the institutions' student and academic affairs; clear learning objectives with a strong academic focus; and flexibility to capitalize on learning opportunities wherever and whenever they occur (Brower and Inkelas, 2010). In a review of the effects of living-learning programs on women seeking a STEM degree (Inkelas et al., 2008), no clear pattern was seen. However, women in STEM-focused programs did rate their residential environments as more academically and socially supportive than women not in those programs, and they rated their sense of belonging and self-confidence higher than did their counterparts.

One example of a STEM-focused living-learning program that illustrates how institutions are implementing such programs is the Living-Learning Community for Women in STEM at the Douglass College of Rutgers University.[20] As part of this program, women studying STEM live in the same residential hall. The residents are provided access to peer study groups, academic and professional development seminars, internship opportunities, roundtable discussions with faculty, and a resource library. In addition, a one-credit course on careers in STEM is required of students in the program. All participants are expected to meet regularly with a graduate mentor and actively participate in learning opportunities in the residence hall.

---

[20]For details, see https://douglass.rutgers.edu/bunting-cobb-residence-hall-living-learning-community-women-stem-0 [April 2015].

Comprehensive Interventions

Programs, such as the Meyerhoff Scholars Program at the University of Maryland Baltimore County, have been lauded for addressing the social and relational aspects of STEM learning. These programs usually provide a range of co-curricular supports to students, as well as implementing changes in classroom instructional practices, changing expectations of faculty for students from underrepresented minorities, and building state-of-the art learning facilities.

The Meyerhoff Scholarship Program began in 1988 with funding from Robert and Jane Meyerhoff and the leadership of then provost (later president) Freeman Hrabowski. Howard Hughes Medical Institute and the National Institutes of Health later provided funding as well. The initial goal of the program was to provide financial assistance, mentoring, advising, and research experience to highly qualified black male undergraduate students committed to obtaining Ph.D. degrees in mathematics, science, and engineering.[21] In 1990, the program was expanded to include black female students, and it was opened up to male and female students of all backgrounds in 1996. According to its website, the program operates on the "premise that, among like-minded students who work closely together, positive energy is contagious. By assembling such a high concentration of high-achieving students in a tightly knit learning community, students continually inspire one another to do more and better."[22]

All incoming Meyerhoff Scholars attend an accelerated 6-week residential program, called summer bridge. The idea of the summer bridge is to teach students about the program and its approach, as well as to provide tools and skills that will help them in their first semester of college. During the summer, students take for-credit courses in calculus and black studies, as well as noncredit courses in chemistry, physics, study skills, and time management. Courses are designed to demonstrate the rigors of college-level instruction and to help students learn how to meet higher standards of performance.

The program focuses heavily on pushing students toward a goal of achieving a Ph.D. The oversight of Meyerhoff Scholars is highly structured, with frequent advising on academics, preparation for graduate and professional school, and assistance with any personal issues that may interfere with school. Students are encouraged to seek not just the A grades, but high–A grades. Advisors, mentors, and peer coaches discuss values, such as outstanding academic achievement, seeking help (tutoring, advising) from a

---

[21] For details, see http://meyerhoff.umbc.edu/about/ [April 2015].

[22] For details, see http://www.umbc.edu/Programs/Meyerhoff/about_the_program.html [April 2015].

variety of sources, and supporting one's peers. Students are told repeatedly that nothing is impossible if they try hard enough.

The program has identified 13 key components to their success: recruitment, financial aid, summer bridge, study groups, program values, program community, tutoring, advising and counseling, professional and faculty mentors, summer research internships, faculty involvement, administrative involvement, and family involvement.[23] All Meyerhoff Scholars are expected to begin participating in research early in their college careers. Since 1993, the program has graduated over 900 students. As of January 2015, the program has achieved the following results:

- Alumni from the program have earned 209 Ph.D.s, which includes 43 M.D./Ph.D.s, 1 D.D.S./Ph.D., and 1 D.V.M./Ph.D. Graduates have also earned 239 master's degrees, as well as 107 M.D. degrees. Meyerhoff graduates have received these degrees from many top institutions, including the University of California at Berkeley, Carnegie Mellon, Duke, Georgia Institute of Technology, Harvard, Johns Hopkins, Massachusetts Institute of Technology, New York University, Rice, Stanford, University of Maryland, University of Michigan, University of Pittsburgh, and Yale.
- More than 300 alumni are currently enrolled in graduate and professional degree programs.
- An additional 270 students were enrolled in the program for the 2015–2016 academic year, of whom 51 percent were black, 15 percent were white, 15 percent were Asian, 12 percent were Hispanic, and 1 percent were Native American.
- Meyerhoff Scholars were 5.3 times more likely to have graduated from or be currently attending a STEM Ph.D. or M.D./Ph.D. program than those students who were invited to join the program but declined and attended another university.

In spring 2014, Howard Hughes Medical Institute agreed that it would fund a 5-year partnership between University of Maryland, Baltimore County; University of North Carolina, Chapel Hill; and the Pennsylvania State University, University Park to help faculty members and administrators document crucial aspects of the program in order to provide guidance for those seeking to replicate it.

---

[23] For more information, see http://meyerhoff.umbc.edu/13-key-components/ [August 2015].

SUMMARY

Students encounter STEM through the environment of a specific department and discipline as reflected in the curriculum, classroom, laboratory, and research experience. They also encounter the environment of STEM through interactions with faculty, staff, and peers, unrelated to instruction, as well as in the expectations, behaviors, and beliefs of those around them. Based on the nature of these interactions, students can be led either to adoption of a STEM identity and to finding and thriving in a STEM community where there is affirmation and support, or they can be pushed into isolation, disaffection, or abandonment of their goals in STEM.

Instructional strategies that have demonstrated efficacy regardless of discipline include more time with students engaged in active learning, and the use of formative assessment and feedback. Significant resources have been invested in disseminating effective practices. There is emerging evidence on the rate of change. Existing evidence makes it difficult to know what percentage of classrooms or departments have adopted effective classroom strategies. However, we do know that the nature of faculty appointments is associated with the learning environment that STEM students encounter. Teaching strategies of part-time contingent faculty are less likely to reflect the qualities of effective instructional strategies, in comparison to tenured or tenure-track faculty. In addition, changes in instructional strategies can be difficult due to a lack of institutional incentives for faculty to change their instructional strategies, minimal time to research and implement evidence-based strategies, and a lack of resources to invest in evidence-based strategies.

Although classroom reform, co-curricular programming, or integrative reforms can address the normative STEM culture that sends negative messages to students, especially to women and those from underrepresented minority groups, about their ability and belonging in the disciplines, students also face barriers to earning a STEM degree that arise from departmental, institutional, and national policies. Awareness of these barriers has become increasingly acute as the ways that students navigate the higher education system have become increasingly complex.

REFERENCES

American Association for the Advancement of Science. (2011). *Vision and Change in Undergraduate Biology Education: A Call to Action.* Final report of a National Conference organized by the American Association for the Advancement of Science, Washington, DC. Available: http://visionandchange.org/files/2011/03/Revised-Vision-and-Change-Final-Report.pdf [April 2015].

American Chemical Society. (2010). *Who Is Teaching Whom? Complete Report on the Fall 2009 CPT Survey of Chemistry Faculty Status.* Washington, DC. Available: https://www. acs.org/content/acs/en/about/governance/committees/training/reports/faculty-status-survey-report.html [April 2015].

American River College. (1993). *A.R.C. Beacon Project: Student Catalyst Program—Peer Assisted Learning: First Semester Summary Report.* Sacramento, CA: American River College.

Annis, L.F. (1983). The process and effects of peer tutoring. *Human Learning, 2*(1), 39–47.

Arco-Tirado, J. L., Fernández-Martín F.D., and Fernández-Balboa, J. (2011). The impact of a peer-tutoring program on quality standards in higher education. *Higher Education, 62*(6), 773–788.

Astin, A.W. (1984). Student involvement: A developmental theory for higher education. *Journal of College Student Personnel, 25,* 297–308.

Austin, A. (2010). Supporting faculty members across their careers. In K. Gillespie and D.L. Robertson (Eds.), *Guide to Faculty Development: Practical Advice, Examples, and Resources* (pp. 363–378). San Francisco: Jossey-Bass.

Austin, A.E. (2011). *Promoting Evidence-Based Change in Undergraduate Science Education.* Paper prepared for the Committee on Barriers and Opportunities in Completing 2-Year and 4-Year STEM Degrees, National Academy of Sciences, Washington, DC. Available: http://sites.nationalacademies.org/cs/groups/dbassesite/documents/webpage/dbasse_072578.pdf [April 2015].

Baldwin, R., and Wawrzynski, M. (2011). Contingent faculty as teachers: What we know; what we need to know. *American Behavioral Scientist, 55*(11), 1485–1509.

Beach, A.L, Henderson, C., and Finkelstein, N. (2012). Facilitating change in undergraduate STEM education. *Change: The Magazine of Higher Learning, 44*(6), 52–59.

Beichner, R.J., Saul, J.M., Abbott, D.S., Morse, J.J., Deardorff, D.L., Allain, R.J., Bonham, S.W., Dancy, M.H., and Risley, J.S. (2007). The Student-Centered Activities for Large Enrollment Undergraduate Programs (SCALE-UP) project. In E. Redish and P.J. Cooney (Eds.), *Research-Based Reform of University Introductory Physics.* Available:http://www. per-central.org/document/ServeFile.cfm?ID=4517 [April 2015].

Benware, C.A., and Deci, E.L. (1984). Quality of learning with an active versus passive motivational set. *American Educational Research Journal, 21,* 755–765.

Borrego, M., and Henderson, C. (2014). Increasing the use of evidence-based teaching in STEM higher education: A comparison of eight change strategies. *Journal of Engineering Education, 103*(2), 220–252.

Borrego, M., Froyd, J.E. and Hall, T.S. (2010). Diffusion of engineering education innovations: A survey of awareness and adoption rates in U.S. engineering departments. *Journal of Engineering Education, 99*(3), 185–207.

Bressoud, D., Carlson, M., Mesa, V., and Rasmussen, C. (2013). The calculus student: Insights from the MAA National Study. *International Journal of Mathematical Education in Science and Technology, 44*(5), 685–698. Available: http://www.researchgate.net/publication/260459087_The_calculus_student_Insights_from_the_MAA_national_study [April 2015].

Brewer, C.A., and Smith, D. (2010). *Vision and Change in Undergraduate Biology Education: A Call to Action.* Washington, DC: American Association for the Advancement of Science.

Brower, A.M., and Inkelas, K.K. (2010). Living-learning programs: On high-impact educational practice we know a lot about. *Liberal Education, 96*(2), 36–43.

Chang, M.J., Eagan, M.K., Lin, M.H., and Hurtado, S. (2011). Considering the impact of racial stigmas and science identity: Persistence among biomedical and behavioral science aspirants. *Journal of Higher Education, 82*(5), 564–596.

Cohen, G.L., Steele, C.M., and Ross, L.D. (1999). The mentor's dilemma: Providing critical feedback across the racial divide. *Personality and Social Psychology Bulletin, 25,* 1302–1318.

Cooper, E. (2010). Tutoring centre effectiveness: The effect of drop-in tutoring. *Journal of College Reading and Learning, 40*(2), 21-34.

Cummings, K., Marx, J., Thornton, R., and Kuhl, D. (1999). Evaluating innovation in studio physics. *American Journal of Physics, 67,* S38–S45.

Dancy, M., and Henderson, C. (2010). Pedagogical practices and instructional change of physics faculty. *American Journal of Physics, Physics, 78*(10), 1056–1063.

Duschl, R.A. (2002). Teaching and learning science: A personalized approach. *Science Education, 86*(5), 727–730.

Eagan, M.K., and Jaeger, A.J. (2008). Closing the gate: Part-time faculty instruction in gatekeeper courses and first-year persistence. In J.M. Braxton (Ed.), *The Role of the Classroom in College Student Persistence: New Directions for Teaching and Learning* (no. 115, pp. 39–53). San Francisco: Jossey-Bass.

Eagan, M.K. (2013). *Understanding undergraduate interventions in STEM: Insights from a national study.* Presented to the Committee on Barriers and Opportunities in Completing 2-Year and 4-Year STEM Degrees, National Academy of Sciences, Washington, DC. Available: http://sites.nationalacademies.org/cs/groups/dbassesite/documents/webpage/dbasse_085900.pdf [July 2015].

Eagan, M.K., Hurtado, S., Chang, M.J., Garcia, G.A., Herrera, F.A., and Garibay, J.C. (2013). Making a difference in science education: The impact of undergraduate research programs. *American Educational Research Journal, 50*(4), 683–713.

Eagan, M.K., Stolzenber, E.B., Berdan Lozano, J., Aragon, M.C., Suchard, M.R., and Hurtado, S. (2014). *Undergraduate Teaching Faculty: The 2013–2014 HERI Faculty Survey.* Los Angeles: Higher Education Research Institute, University of California.

Eddy, S.L., and Hogan, K.A. (2014). Getting under the hood: How and for whom does increasing course structure work? *Life Science Education, 13,* 453–468.

Estrada, M. (2014). *Ingredients for Improving the Culture of STEM Degree Attainment with Co-curricular Supports for Underrepresented Minority Students.* Paper prepared for the Committee on Barriers and Opportunities in Completing 2-Year and 4-Year STEM Degrees, National Academy of Sciences Washington, DC. Available: http://sites.nationalacademies.org/cs/groups/dbassesite/documents/webpage/dbasse_088832.pdf [April 2015].

Fairweather, J. (2008). *Linking Evidence and Promising Practices in Science, Technology, Engineering, and Mathematics (STEM) Undergraduate Education: A Status Report for the National Academies National Research Council Board on Science Education.* Paper prepared for the Workshop on Evidence on Promising Practices in Undergraduate Science, Technology, Engineering, and Mathematics (STEM) Education. Available: http://www.nsf.gov/attachments/117803/public/Xc--Linking_Evidence--Fairweather.pdf [April 2015].

Felder, R.M., and Brent, R. (2010). The National Effective Teaching Institute: Assessment of impact and implications for faculty development. *Journal of Engineering Education, 99*(2), 121–134.

Felder, R.M., Brent, R., and Prince, M.J. (2011). Engineering instructional development: Programs, best practices, and recommendation. *Journal of Engineering Education, 100*(1), 89-122.

Fifolt, M., and Searby, L. (2010). Mentoring in cooperative education and internships: Preparing proteges for STEM professions. *Journal of STEM Education, 11*(1&2), 17–26.

Freeman, S., Eddy. S.L., McDonough, M., Smith, M.K., Okoroafor, N., Jordt, H., and Wenderoth, M.P. (2014). End of lecture: Active learning increases student performance across the STEM disciplines. *Proceedings of the National Academy of Sciences of the United States of America, 111*(23), 8410–8415.

Gaffney, J.D.H, Richards, E., Kustusch, M.B., Ding, L., and Beichner, R. (2008). Scaling up educational reform. *Journal of College Science Teaching, 37*(5), 48–53.

Gandara, P., and Maxwell-Jolly, J. (1999). *Priming the Pump: Strategies for Increasing Achievement of Underrepresented Minority Graduates.* New York: The College Board.

Gilmer, T.C. (2007). An understanding of the improved grades, retention and graduation rates of STEM majors at the Academic Investment in Math and Science (AIMS) Program of Bowling Green State University (BGSU). *Journal of STEM Education, 8*(1), 11–21.

Henderson, C., and Dancy, M. (2007) Barriers to the use of research-based instructional strategies: The influence of both individual and situational characteristics. *Physical Review Special Topics: Physics Education Research, 3*(2). Available: http://journals.aps.org/prstper/abstract/10.1103/PhysRevSTPER.3.020102 [April 2015].

Henderson, C. (2008) Promoting instructional change in new faculty: An evaluation of the physics and astronomy new faculty workshop. *American Journal of Physics, 76*(2), 179–187.

Henderson, C., and Dancy, M. (2009). The impact of physics education research on the teaching of introductory quantitative physics in the United States. *Physics Review Special Topics: Physics Education Research, 5*(2), 020107.

Henderson, C., Beach, A., and Finkelstein, N. (2011). Facilitating change in undergraduate STEM instructional practices: An analytic review of the literature. *Journal of Research in Science Teaching, 48*(8), 952–984.

Henderson, C., Dancy, M., and Niewiadomska-Bugaj, M. (2012) The use of research-based instructional strategies in introductory physics: Where do faculty leave the innovation-decision process? *Physical Review Special Topics: Physics Education Research, 8*(2). Available: http://journals.aps.org/prstper/abstract/10.1103/PhysRevSTPER.8.020104 [April 2015].

Hilborn, R.C., and Howes, R.H. (2003). Why many undergraduate physics programs are good but few are great. *Physics Today, 56*(9), 38–45.

Hilborn, R.C. (2013). *The Role of Scientific Societies in STEM Faculty Workshops: A Report of the May 3, 2012, Meeting.* Available: https://www.aapt.org/Conferences/newfaculty/upload/CSSP_May_3_Report_Final_Final_Version_3-15-13.pdf [April 2016].

Hoellwarth, C.C., Moelter, M.J., and Knight, R. (2005). A direct comparison of conceptual learning and problem-solving ability in traditional and studio style classrooms. *American Journal of Physics, 73*(5), 459–462. Available: http://works.bepress.com/choellwa/1 [April 2015].

Hurtado, S., Cabrera, N.L., Lin, M.H., Arellano, L., and Espinosa, L.L. (2009). Diversifying science: Underrepresented student experiences in structured research programs. *Research in Higher Education, 50*, 189–214.

Hurtado, S., Eagan, M.K., Tran, M.C., Newman, C.B., Chang, M.J., and Velasco, P. (2011). We do science here: Underrepresented students' interactions with faculty in different college contexts. *Journal of Social Issues, 67*(3), 553–579.

Inkelas, K.K., and Associates. (2008). *The National Study of Living-Learning Programs: 2007 Report of Findings.* Available:http://drum.lib.umd.edu/bitstream/1903/8392/1/2007%20NSLLP%20Final%20Report.pdf [April 2015].

Jaeger, A.J., and Eagan, M.K. (2009). Effects of exposure to part-time faculty on associate's degree completion. *Community College Review, 36*, 167–194.

Jaeger, A.E., and Eagan, M.K. (2011). Navigating the transfer process: Analyzing the effects of part-time faculty exposure by academic program. *American Behavioral Scientist, 55*(11), 1510–1532.

Kinkead, J. (2003). Learning through inquiry: An overview of undergraduate research. *New Directions for Teaching and Learning, 93*, 5–17.

Ko, L.T., Kachchaf, R.R., Hodari, A.K., and Ong, M. (2014). Agency of women of color in physics and astronomy: Strategies for persistence and success. *Journal of Women and Minorities in Science and Engineering, 20*(2), 171–195.

Kuh, G.D. (2008). Excerpt from "High-Impact Educational Practices: What They Are, Who Has Access to Them, and Why They Matter." Available: https://www.aacu.org/leap/hips [April 2015].

Lenaburg, L., Aguirre, O., Goodchild, F., and Kuhn, J.U. (2012). Expanding pathways: A summer bridge program for community college STEM students. *Community College Journal of Research and Practice, 36*(3) 153–168.

Lidren, D.M., Meier, S.E., and Brigham, T.A. (1991). The effects of minimal and maximal peer tutoring systems on the academic performance of college students. *Psychological Record, 41*(1), 69–77.

Lopatto, D. (2003). The essential features of undergraduate research. *Council on Undergraduate Research Quarterly,* (2), 139–142.

Mabrouk, P.A., and Peters, K. (2000). Student perspectives on undergraduate research (UR) experiences in chemistry and biology. *Council on Undergraduate Research Quarterly, 21*, 25–33.

Macdonald, R.H., Manduca, C.A., Mogk, D.W., and Tewksbury, B.J. (2005). Teaching methods in undergraduate geoscience courses: Results of the 2004 Cutting Edge Survey of U.S. Faculty. *Journal of Geoscience Education, 53*(3), 237–252.

Mason, B. (2007). *Introducing ComPADRE. Forum on Education. 2007. Communities for Physics and Astronomy Digital Resources in Education.* Available: http://www.compadre.org [April 2015].

Nadal, K., Wong, Y., Grifin, K.E., Davidoff, K., and Sriken, J. (2014). The adverse impact of racial microaggressions on college students' self-esteem. *Journal of College Student Development, 55*(5), 461–474.

National Research Council. (2012). *Discipline-Based Education Research: Understanding and Improving Learning in Undergraduate Science and Engineering.* Committee on the Status, Contributions, and Future Directions of Discipline-Based Education Research. S. Singer, N.R. Nielsen, and H.A. Schweingruber (Eds.). Board on Science Education, Division of Behavioral and Social Sciences and Education. Washington, DC: The National Academies Press.

National Research Council. (2013). *Adapting to a Changing World: Challenges and Opportunities in Undergraduate Physics Education.* Committee on Undergraduate Physics Education Research and Implementation. Board on Physics and Astronomy. Division on Engineering and Physical Sciences. Washington, DC: The National Academies Press.

Pascarella, E.T., and Terenzini, P.T. (2005). *How College Affects Students, Volume 2, A Third Decade of Research.* San Francisco, CA: Jossey-Bass.

Pike, G.R. (1999). The effects of residential learning communities and traditional residential living arrangements on educational gains during the first year of college. *Journal of College Student Development, 40*, 269–284.

Pike, G.R., Schroeder, C.C., and Berry, T.R. (1997). Enhancing the educational impact of residence halls: The relationship between residential learning communities and first year college experiences and persistence. *Journal of College Student Development, 38*, 609–621.

President's Council of Advisors on Science and Technology. (2012). *Report to the President. Engage to Excel: Producing One Million Additional College Graduates with Degrees in Science, Technology, Engineering and Mathematics.* Available: http://www.whitehouse. gov/sites/default/files/microsites/ostp/pcast-engage-to-excel-final_feb.pdf [April 2015].

Prince, M., Borrego, M., Henderson, C., Cutler, S., and Froyd, J. (2013) Use of research-based instructional strategies in core chemical engineering courses. *Chemical Engineering Education, 47*(1), 27–37.

Savkar, V., and Lokere, J. (2010). *Time to Decide: The Ambivalence of the World of Science Toward Education.* Nature Education position paper. Available: http://i.zdnet.com/blogs/ time-to-decide-nature-education-report-1.pdf [April 2016].

Seymour, E. (2001). Tracking the process of change in U.S. undergraduate education in science, mathematics, engineering, and technology. *Science Education, 86,* 79–105.

Shapiro, N.A., and Levine, J.J. (1999). *Creating Learning Communities: A Practical Guide to Winning Support, Organizing for Change, and Implementing Programs.* San Francisco, CA: Jossey-Bass.

Schunk, D.H. (1987). Peer models and children's behavioral change. *Review of Educational Research, 57,* 149–174.

Sorensen, C.M., Churukian, A.D., Maleki, S., and Zollman, D.A. (2006). The New Studio format for instruction of introductory physics. *American Journal of Physics, 74*(12), 1077–1082.

Stassen, M.L.A. (2003). Student outcomes: The impact of varying living-learning community models. *Research in Higher Education, 44,* 581–613.

Strayhorn, T.L. (2010b), Undergraduate research participation and STEM graduate degree aspirations among students of color. *New Directions for Institutional Research,* 85–93.

Thiry, H., Laursen, S.L., and Hunter, A.B. (2011). What experiences help students become scientists? A comparative study of research and other sources of personal and professional gains for STEM undergraduates. *Journal of Higher Education, 82*(4), 357–388.

Titus-Becker, K., Rajala, S.A., Bottomley, L., Raubenheimer, D., Cohen, J.A., Bullett, K., Gant, S., and Thomas, C. (2007). *An Integrated Living and Learning Community for First- and Second-Year Undergraduate Women in Science and Engineering.* Paper presented at the American Society of Engineering Education Annual Conference and Exposition. Available: https://peer.asee.org/an-integrated-living-and-learning-community-for-first- and-second-year-undergraduate-women-in-science-and-engineering [October 2015].

Topping, K.J. (1996). The effectiveness of peer tutoring in further and higher education: A typology and review. *Higher Education, 32*(3), 321–345.

Umbach, P., and Wawrzynski, M. (2005). Faculty do matter: The role of college faculty in student learning and engagement. *Research in Higher Education, 46*(2), 153–184.

Weaver, G.C., Russell, C.B., and Wink, D.J. (2008). Inquiry-based and research-based laboratory pedagogies in undergraduate science. *Nature Chemical Biology, 4*(10), 577–580.

5

Institutional, State, and National Policies

Major Messages

- Some institutional, state, and national undergraduate education policies present significant barriers to students' progress toward a science, technology, engineering, and mathematics (STEM) credential.
- On average, the price a student pays for a STEM credential is higher than the price of a non-STEM credential, reflecting the relatively higher cost to institutions of producing a STEM credential, compared with a non-STEM credential.
- Academic departments can smooth students' pathways to a STEM credential by developing inter-institutional agreements that simplify the transfer process and support transfer students, reducing course sequencing restrictions, reducing degree requirements and prerequisites outside of the major, adjusting grading practices, and adopting creative solutions to improve and reduce the need for remedial courses.

Institutional, state, and national policies can become significant barriers for students as they follow complex pathways to earn STEM credentials (including degrees and certificates). Given the current pattern of frequent transferring between institutions and in earning credits at multiple institutions (either at the same time or sequentially; Hossler et al., 2012), policies related to transferring credits are key factors in students' progress toward a credential. Related to this factor is the cost of earning a credential. In this chapter, we discuss the institutional and systemic factors that affect transfer policies and costs.

Since transfer policies and degree costs are connected to institutional, state, and national policies and practices that affect all students, not just those in STEM fields, this chapter includes discussion of the broader policy context. When data are available, we discuss STEM-specific aspects of transfer and degree cost and the role that STEM departments can play in addressing issues related to these policies.

BARRIERS ASSOCIATED WITH TRANSFERRING

An undergraduate's likelihood of completing a bachelor's degree is much lower if the student begins at a community college than at a 4-year institution (Alfonso, 2006; Monaghan and Attewell, 2014; Reynolds, 2012; Reynolds and DesJardins, 2009; Stephan et al., 2009). Many academically qualified students in community colleges who intend to transfer and earn a bachelor's degree never do so. Although there is limited evidence about what prevents students from transferring, contributing factors may be lack of appropriate advice, conflicting personal commitments, and geographic accessibility (Monaghan and Attewell, 2014). However, those community college students who are able to transfer to 4-year institutions are just as likely to complete a bachelor's degree as students who initially enrolled at 4-year institutions (Melguizo et al., 2011). Some of this difference in bachelor's degree attainment by students who first enroll in a community college is due to the entry characteristics of community college students, including academic preparation, socioeconomic status, demographic background, and levels of parental education. The increased likelihood of part-time status at a community college also negatively affects graduation rates (The College Board, 2014).

The difference in degree completion rates between students who start at a community college and those who start at a 4-year institution has repercussions with respect to the goal of increasing the diversity of college graduates. Until this difference is reduced, it may be difficult to increase the representation of underrepresented minority groups and those from low-income families among college graduates, as these student groups are more likely to enroll at a community college (Bozick and Lauff, 2007; National Center for Public Policy and Higher Education, 2011). Looking at enrolled postsecondary students in 2012, 46 percent of Hispanic, 41 percent of American Indian/Alaska Native, and 35 percent of black college students were enrolled at a community college in comparison with 31 percent of white students (National Center for Education Statistics, 2013).

Similarly, students who reverse transfer, from 4-year to 2-year institutions, are less likely to earn a bachelor's degree than those who transfer to another 4-year institution: only 22 percent of reverse transfer students earn a bachelor's degree within 8 years of their initial enrollment. Reverse transfer is most common among students who struggled academically in their first year and those whose parents have less education than otherwise comparable students (Goldrick-Rab, 2006).

"Transfer shock" is the term coined in the mid-1960s (Hills, 1965) to refer to the tendency for there to be a temporary dip (typically only lasting one semester) in the grade point averages of students who transfer—a trend that still exists today. There is no clear single reason for the dip. Some early research (e.g., Townsend, 1995) pointed to the tendency for transfer students to seek support from friends and family, rather than from on-campus support services. They also proposed that transfer students often find the 4-year institution to have higher academic standards, be faster paced, and require a greater amount of writing than did their community colleges (Townsend, 1995). Other researchers (Holahan et al., 1983; Laanan, 1996, 1998) have linked transfer students' difficulties adjusting to the academic standards of 4-year institutions to institutional differences in size, location, academic rigor, and competition among students. Other studies have looked at social and psychological factors that might contribute to transfer shock. In a review of the literature on transfer shock, Laanan (2001) points to differences in campus climate and the accessibility of faculty.

The intertwined effects of time to degree and cost of degree also contribute to lower completion rates for transfer students. The longer students extend their enrollment in college beyond their expected graduation date (i.e., beyond 4 years for a 4-year degree), the lower are the odds that they will graduate (Complete College America, 2011). Monaghan and Attwell (2014) found that many students transferring from community colleges lose credits because the credits are not approved by the 4-year institution. In their study sample, 14 percent of transfer students had less than 10 percent of their credits accepted, and only 58 percent of transfer students had more than 90 percent of their credits accepted. As the percentage of community college credits transferred increased, the likelihood of attaining a bachelor's degree also increased (Monaghan and Attewell, 2014). In a separate study, Doyle (2006) found that when 4-year institutions accepted all of the community college credits, 87 percent of transfer students earned a bachelor's degree. In addition, this study found that when many credits, but not all were accepted, the percentage of transfer students who earned a degree in the same amount of time dropped to 42 percent.

Institutions may be hesitant to accept transfer credits for a number of reasons. They may question the rigor of the coursework at the institution from which the credit is being transferred, and they may question the level of alignment between the content of courses across institutions. In addition, institutions have begun to question how many credits a student needs at one institution to earn a degree. For example, should students who complete their degree requirements from university X, while only earning 40 percent (or less) of their credits from that university, be granted a degree from university X? What does it then mean to earn a degree from university X? However, even acceptance of transfer credit does not necessarily translate into fewer credits needed to earn a degree. Degree requirements typically include courses in three categories: (1) general education, (2) courses specific to a major, and (3) elective credits. If transfer credits are classified as elective and exceed the credit requirements, then the additional credits will not contribute to the degree and are effectively lost, even though they have transferred. It has been estimated that the cost of credits that do not advance students toward their degrees is over $7 billion per year (Smith, 2010).

TRANSFER POLICIES AND THEIR EFFECTS

Most colleges and universities are accredited by one of the seven U.S. regional accreditors of higher education; a process requires them to demonstrate compliance with standards that include published and implemented policies, procedures, and criteria regarding transfer credit and credit for extra-institutional college-level learning. Accreditation is required in order for a college's students to be eligible for federal financial aid. Articulation agreements are a mechanism for formalizing a college's transfer credit policies and are often designed to create a clear pathway for transfer, promote appropriate preparation for future academic work, encourage vertical transfer (from 2-year to 4-year institutions), and maximize credits transferred, among other goals.[1] As of 2014, only five states rely solely on institutional policies for articulation—Delaware, Iowa, Michigan, Vermont, and Wyoming. The other 45 states have some statewide and systemwide transfer policies, although they vary greatly (Western Interstate Commission for Higher Education, 2014).

The number of states with transfer policies has been steadily increasing over the last few decades. In 1991, only 12 states had adopted statewide policies (Anderson et al., 2006); by 1999, the number had increased to 34 (Goldhaber et al., 2008). As shown in Figure 5-1, by 2010, over 35 states had developed formal statewide transfer policies. These policies can take the form of laws, or legislative recommendations, or they can be set by state boards (Western Interstate Commission for Higher Education,

---

[1] Articulation (or course articulation) is the comparison of courses between institutions in order to determine their equivalency for purposes of transfer credit.

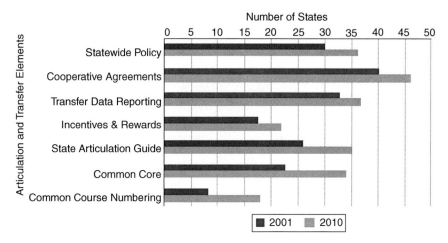

FIGURE 5-1 Number of states with articulation and transfer elements, 2001 and 2010.
SOURCE: Mullin (2012, Figure 1).

2014). Components often include a common core of general education requirements, common major-specific courses, common course numbering systems, guaranteed junior status for students who earned a 2-year degree from a community college, online access to transfer and degree program information, upper limits on the number of required credits for associate or baccalaureate degrees, and guaranteed or priority admission for transfer students (Kisker et al., 2011). Transfer equivalents are determined on a course-by-course basis or based on common learning outcomes.

Early studies of the relationship between transfer policies and student aspirations and outcomes failed to reach conclusive results (Roksa, 2009; Roksa and Keith, 2008). Goldhaber and colleagues (2008) found a positive, but inconsistent correlation between stronger transfer policies and increases in community college students' educational aspirations and in the number of students transferring from 2-year to 4-year institutions. However, they did not consistently find that the states with statewide policies or stronger policy elements had the highest share of community college students aspiring to a 4-year degree. In a subsequent analysis, Gross and Goldhaber (2009) failed to find that any stronger policy elements had statistically significant effects on the likelihood of 2-year students earning a bachelor's degree.

A recent study found that the higher students' educational aspirations were, the more likely they were to earn a bachelor's degree (Monaghan and Attewell, 2014). This study also found that the percentage of students who earned a 4-year degree was slightly higher in states with transfer agreements than in those without (56% and 51%, respectively). Similarly, among students who reverse transferred, more earned an associate's degree in states with formal agreements than in states without (22% and 16%, respectively).[2]

A review of statewide or systemwide policies (Kisker et al., 2011) considered seven elements in four states: Arizona, New Jersey, Ohio, and Washington:

1. a common general education core curriculum,
2. common lower-division courses in the major,
3. a focus on credit applicability,
4. junior status upon transfer for associate's degree recipients,
5. guaranteed or priority admission,
6. limits on credit requirements for both associate's and bachelor's degrees, and
7. an acceptance policy for upper-division courses that results in what is referred to as transfer associate's degrees.

The authors concluded that the first four elements were critical for beneficial student outcomes in those states.

Community college students in those states had greater flexibility and options than in other states. Transferring to any of the participating colleges within the same degree program was easier for students than when articulation agreements were reached on an institution-to-institution basis. Transfer rates improved in Ohio and Washington. In Ohio, vertical transfers increased by 21 percent between 2002 (the year before common lower-division premajor and early major pathways were first introduced) and 2009, while enrollment only grew by 7 percent.

In addition, the review revealed that policies in Arizona and Ohio were associated with increases in community college students' preparation for upper-division work. For example, in Arizona, students who completed either the state's 35-credit general education common core (called the Arizona General Education Curriculum) or a full transfer associate's degree prior to transfer had significantly higher grade point averages after two and four semesters at the 4-year institution in comparison with students who transferred without completion of either a degree or the general education common core.

Degree completion rates increased in Ohio and Washington, particu-

---

[2]For their study, Monaghan and Attewell (2014) used data from the National Education Longitudinal Survey, 1988–2000 (NELS88/2000) with the NELS 2000 follow-up and the 1999 Survey of State-Level Transfer and Articulation Policies conducted by Ignash and Townsend (2001).

larly for underrepresented groups. For example, Hispanic students in Washington who participated in the transfer associate's degree programs had a particularly high rate of 93 percent. In Arizona and Washington, the time to degree was reduced. In Washington State, students with transfer associate's degrees in science or engineering earned a bachelor's degree with 6 fewer credits than those who completed only the general education common core, and with 49 fewer credits than students who had completed a technical or more traditional associate's degree before transferring. Finally, a separate analysis by the Ohio Board of Regents (Mustafa et al., 2010) found that transfer activities saved the state $20 million per year, of which $7 million was attributed to transfer associate's degrees.

The available research indicates that statewide transfer policies, including a common general education core and transfer associate's degrees, represent an opportunity for increasing the number of students who complete degrees at 4-year institutions and for other positive outcomes as well. A roadmap of sorts is thus available to policy makers in other states who wish to pursue programs to help transfer students and save money.

Recently, some regional accreditors have permitted a competency-based model for earning college credits. The Council of Regional Accrediting Commissions (2015) issued a common framework for defining and accrediting competency-based education.[3] Some states and institutions believe the process of transferring credits could be improved by shifting to a model of learning outcomes or student competencies. For example, if a 4-year institution finds that a course completed by a community college student is missing a minor, required component, the student could complete a short learning module and an assessment to demonstrate knowledge of this missing competency. This competency-based approach could help reduce a student's time to degree and cost of degree by avoiding repetition of a course in which most of the material has already been mastered. More evidence on competency-based models is needed in order to ascertain the impact on transferring credits and degree completion.

As described in Chapter 3, the departmental environment can play an important role in the experiences of students in STEM majors. Not only can departments influence faculty reward systems, course sequence, teaching practices, and departmental culture, but they can also have an effect on transfer policies. While statewide transfer policies and support structures can promote the transfer of credits (see below), departmental leaders ultimately decide on whether or not to accept transfer credits in a discipline (Austin, 2011). In making these decisions, a focus on course learning outcomes, rather than strictly on content coverage, can increase the number

---

[3] See https://www.insidehighered.com/sites/default/server_files/files/C-RAC%20CBE%20Statement%20Press%20Release%206_2.pdf [October 2015].

BOX 5-1
Departmental Policies to Support Transfers: An Example

The Dual Admission Program (DAP) at Florida International University (FIU), which depends on departmental transfer policies, has shown some success in supporting transfer students seeking a STEM degree. The program's primary mission is to increase bachelor's degree attainment by capitalizing on the existing collaborative relationships between FIU and its local and regional transfer feeder institutions. DAP offers local and regional students who aspire to 4-year degrees but who do not meet the admission criteria for FIU the opportunity to join DAP and be guaranteed admission to FIU at a future time.

The goals of DAP are to:

- increase postsecondary educational opportunities for students and improve their chances of attending a 4-year institution and earning bachelor's degrees;
- support and enhance Florida's 2 + 2 articulation model;
- strengthen relationships with partner colleges; and
- manage and track enrollment patterns among partner colleges.

Students agree to matriculate at one of four partner colleges—Broward College, Florida Keys Community College, Miami Dade College, and Palm Beach State College—and to complete the associate's degree within 3 years. At that time, students' transition to FIU is done through a reactivation of their admission application. The transition to FIU is streamlined by Florida's Statewide Articulation Agreements and enforcement of common course numbering. DAP students remain affiliated with FIU while attending the partner college: they receive an FIU student ID, have access to targeted FIU resources, including academic advising;

---

of credits accepted for transfer. One challenge is to engage faculty at 4-year institutions in the process, since some think community college students should find their own paths into the disciplinary major at the 4-year institution. The department chair can play an important role in involving faculty members to actively help transfer students (Parker et al., 2014).

Some departments are redesigning academic programs and student services to create more structured paths designed to guide students to transfer with junior standing in their majors—and earn an associate's degree along the way: see Box 5-1 for an example.

and are invited to affinity-building events (e.g., artistic performances, lectures, athletic games). Since it was first launched in 2006, FIU has added features to DAP, which now also includes a bridge program for all students transferring from Miami Dade College (75% of DAP participants initially enroll at MDC), and a host of supports for students who initially enroll at the other partner colleges.

Interviews conducted by FIU with approximately 100 DAP students revealed the following primary themes: (1) a "safe start" at Miami Dade College, which increased their academic confidence to transition to FIU; (2) less financial burden; (3) motivation to attend full time given the stipulated time limit; (4) promise of a "seat" at FIU; and (5) the "feel" of the program as one continuous experience rather than as two distinct institutions. Overall, students took pride in identifying themselves as DAP students attending Miami Dade College. Finally, students who worked with advisors noted that the services provided were filling information and transition gaps.

Quantitative data collected by FIU from the 2006–2007 cohorts—who transitioned to FIU in 2009 and 2010—revealed that 50 percent of DAP students earned an associate's degree within a 3-year period (673 students), in comparison with 18 percent for the non-DAP students (257 students). DAP students also completed their associate's degrees in less time (7.7 semesters) than their non-DAP counterparts (8.6 semesters). Interestingly, 72 percent of the DAP participants required at least one developmental course in comparison with 33 percent of students who declined the DAP invitation, but who still attended Miami Dade College. Taken together, these data suggest that students who accepted the invitation to DAP, although needing developmental coursework, appear to be highly motivated and to maintain greater momentum to degree attainment than their counterparts who were not in DAP, though the latter were initially better prepared.

SOURCE: Based on material developed by Elizabeth Bejar and Mark Rosenberg at Florida International University.

INSTITUTIONAL PROGRAMS TO FACILITATE TRANSFERS AND ASSIST STEM STUDENTS

Although most professionals in higher education would agree that forging partnerships between 2-year and 4-year institutions is valuable to smooth the transition for transfer students and enhance their opportunities for success are important goals, actual implementation may prove challenging. Challenges include the different cultures and missions of 2-year and 4-year colleges, perceived competition for students, and geographic separation. When they are successfully established, such partnerships often create articulation agreements, and coordinate a wide range of activities, including advisements; opportunities for community college students to take courses, attend seminars, perform undergraduate research, and otherwise participate in activities at the 4-year institution; arranging for the 4-year institution to offer classes on the campus of the 2-year institution; and even satellite campuses, where a 4-year institution offers degrees and has permanent faculty and staff located on the campus of the 2-year institution.

The College Board (2011), based on interviews of 21 campus leaders from 12 4-year institutions, recommended that institutions should recognize and embrace the contributions that transfer students make to their educational, economic, and cultural diversity and include support of transfer students in their strategic plans. While recognizing that the needs of transfer students are somewhat different than those of first-year students, the College Board (2011) study encouraged 4-year institutions to provide transfer students with the kind of dedicated support that they typically give to first-year students. That support could include outreach prior to admission to promote better preparation for upper-level study and to create a clear transfer pathway; aligning the curricula; dedicating financial aid to transfer students; and offering a dedicated transfer orientation. The report also encouraged 4-year institutions to make work-study positions available to transfer students.

One example of a college that is taking steps to smooth the transfer process is the Onondaga Community College Regional Higher Education Center in New York, which hosts numerous colleges offering bachelor's and master's degree programs.[4] Some of the 4-year colleges that partner with Onondaga Community College offer joint degree programs. These programs allow community college students to simultaneously enroll in an associate's degree and a bachelor's degree program, and once they complete their associate's degree, they attend the 4-year institution. One example is a dual-joint degree program between the John Jay College of Criminal Justice and Queensborough Community College, which offer joint registration and dual admission.[5]

Another model is the dual degree program. One example is the partnership between Governors State University and eight community colleges in the Chicago area. This dual degree program, which enrolled its first cohort in spring 2011, requires students to complete their associate's degrees, attend college full-time, maintain good academic standing, meet regularly with community college and university advisors, and finish both the associate's and bachelor's degrees in no more than nine semesters. Community college students are eligible for guaranteed admission at Governors State University, where they receive academic support and financial incentives.[6]

---

[4]For more details, see http://www.sunyocc.edu/index.aspx?menu=851&collgrid=510&id=29053 [April 2015].

[5]See http://www.qcc.cuny.edu/socialSciences/criminalJustice.html [April 2015].

[6]See http://www.govst.edu/Academics/Degree_Programs_and_Certifications/Dual_Degree_Program/ [April 2015].

BOX 5-2
Research Experiences for Community College Students:
An Example

The Science, Technology, Engineering and Mathematics Talent Expansion Program-University Partnership (NU STEP-UP) is a partnership between Northeastern University, two research centers supported by the National Science Foundation, and three Boston-area community colleges (Massachusetts Bay Community College, Middlesex Community College, and Northern Essex Community College) to increase the number of students receiving degrees in STEM disciplines. NU STEP-UP is focused on developing a sustainable STEM model that provides a seamless transition between 2-year and 4-year institutions. Using research as the catalyst for engagement, NU STEP-UP is (1) creating a sustainable STEM partnership between the university's STEM departments and local community colleges; (2) creating a Partner Faculty Network, with representatives from all stakeholders; (3) providing community college faculty the opportunity to immerse themselves in a research environment; (4) providing community college students access to extensive research experiences; (5) developing a transfer bridge program for community college students transitioning to Northeastern University; and (6) providing academic mentoring and research activities for all STEM students throughout the partnership.

Participants in the Partner Faculty Network are involved in working seminars, helping them implement the latest pedagogical approaches in their own classrooms. They are sharing innovative STEM instructional models and practice and collaborating to bring STEM courses at community colleges in alignment with comparable courses at 4-year institutions. A multifaceted approach to program evaluation aims to assess progress toward achieving established benchmarks, as well as to understand the contribution of various program elements. The evaluation plan includes (1) tracking student transfer rates, retention rates, and student performance; (2) surveys of stakeholders, including students, faculty, and alumni; (3) focus groups with transfer students and with faculty; and (4) cohort analysis of transfer students. Results and outcomes are being disseminated through publications, a project website, and presentations at regional and national conferences.

SOURCE: Information from the project's website, see https://stem-central.net/projects/8#. VYhi2flVhBc [August 2015].

Alliances between community colleges and research universities can also enhance the availability and quality of research experiences for students at community colleges (Shaffer et al., 2010; Wei and Woodin, 2011): for an example, see Box 5-2.

BOX 5-3
Support for Students Who Transfer to Study STEM:
An Example

A program that has shown indications of being successful in supporting transfer students is part of the College of Science at Texas A&M University: the Transfer Learning Community Program. It was developed from a National Science Foundation (NSF) grant program—Scholarships in Science, Technology, Engineering, and Mathematics (S-STEM).

The pilot program targeted 24 students at Palo Alto College, a 2-year institution that primarily served a Hispanic population. Of the students in the program, 21 (88%) successfully transferred to Texas A&M. Of the students who transferred, 18 (86%) had graduated or were on track to graduate after 2 years. The percentage of transfer students in the pilot program who graduated or were on track to graduate was similar among Hispanic students (83%) and non-Hispanic students (89%).

Based on the results of this pilot, elements of the S-STEM program have been institutionalized and scaled up for all incoming students in biology, chemistry, mathematics, and physics, a total of approximately 120 students per year. This semester-long program is intended to help transfer students transition to the university and increase both retention and graduation rates.

Academic Boot Camp is the first required meeting for incoming transfer students. This 3-hour program occurs the Friday prior to the beginning of the fall and spring semesters. Peer mentors are successful transfer students with grade point averages of 3.5 or higher who lead each of the three 1-hour programs describing

---

FEDERAL AND STATE PROGRAMS TO FACILITATE TRANSFRS AND ASSIST STEM STUDENTS

There are a number of federal and state programs that support efforts to promote the success of STEM students who transfer among institutions. For example, the National Science Foundation's S-STEM program supports an innovative transfer program at Texas A&M University (see Box 5-3 for more details). Collaborations were supported by the Career Pathways Innovation Fund of the U.S. Department of Labor (DOL)[7] and through expansion of the mission of the DOL's Trade Adjustment Assistance Community College and Career Training initiative (TAACCCT).[8] These DOL programs encouraged partnerships between community and technical colleges and employers in the private sector to develop scientific research and engineering design exchanges across 2-year and 4-year institutions. TAACCCT

---

[7] See http://www.doleta.gov/grants/pdf/sga-dfa-py-10-06.pdf [April 2015].
[8] See http://www.doleta.gov/taaccct/ [April 2015].

their experiences: the first session focuses on transition to the university, goal setting, and time management; the second session focuses on structured study time, the importance of class attendance, active engagement inside and outside of class, and utilization of campus resources; the final session focuses on reading syllabi, with emphasis on key points, and preparing for exams. Following the presentations, students receive instruction about critical learning theory on each of the above topics, followed by hands-on activities to build their weekly study schedule, discover their learning style, and discuss campus resources with peer mentors.

Following Boot Camp, students are required to attend three 1-hour meetings every month. In the first meeting, peer mentors lead discussion groups about challenges and successes. The purpose of this exercise is to link struggling students with other students who are experiencing success. Successful study strategies are shared, and study groups for different majors are formed. The last portion of the meeting focuses on identifying opportunities to participate in research labs.

The second meeting concentrates on the students' mid-term grades and an understanding of academic standing and advising processes (dropping classes, withdrawing, etc.). During the final session, preregistration for the coming semester is discussed, as is preparing for final exams. Peer mentors lead focus groups to determine issues deemed to be most critical for successful transition to the university. Retention rates rose 2 percent after each of the first 2 years of the program. Grades for the majority of the students also improved during those 2 years.

SOURCE: Adapted from Scott et al. (in press).

grants (awarded from 2011 to 2014) supported a broad set of reforms to improve the infrastructure of community college workforce programs. Many grantees decided to drive reform by improving career pathways including better transfer arrangements. The President's fiscal 2015 budget proposed to establish a Community College Job-Driven Training Fund as a successor to TAACCCT.

Similarly, an important element of the Advanced Technological Education (ATE) Program[9] of the National Science Foundation (NSF) is the development of partnerships between 2-year and 4-year colleges—as well as among secondary schools, with some combination of businesses, industry, and governments—in order to train technicians to meet current and future workplace needs. In 2012, ATE projects and centers supported development of articulation agreements that helped 2,410 students transfer between 2-year and 4-year institutions. The NSF's Louis Stokes Alliances

---

[9] See http://www.nsf.gov/funding/pgm_summ.jsp?pims_id=5464 [April 2015].

for Minority Participation Bridge to the Baccalaureate Alliances[10] Program also supports connections between 2-year and 4-year institutions in order to accelerate the transfer of underrepresented minority STEM degree aspirants to 4-year institutions.

The National Institute of Health's Bridges to the Baccalaureate (BTB) Program is designed to enhance the pool of community college students from underrepresented groups in biomedical and behavioral sciences, with the hope that some of them will go on to research careers in these fields. The program supports implementation of integrated plans to increase community college students' preparation, motivation, and skills and to increase the pool of students who successfully transition from 2-year to 4-year institutions and graduate.

The University of California and the California State University systems have a long-standing program, Mathematics, Engineering, Science Achievement,[11] which creates partnerships between 2-year and 4-year campuses to align curricula and prepare students for the transition to bachelor's degree programs in STEM disciplines. Some other large state systems have similar programs.

There are also programs designed specifically for veterans. A recent study by the American Council on Education (2014) found that veterans currently represent about 4 percent of all undergraduates. Of those veterans, 38 percent attend 2-year public institutions, 23 percent attend private for-profit institutions, 19 percent attend 4-year public institutions, and 10 percent attend private nonprofit institutions. Of these veterans, 42 percent work full time, and 20 percent are in STEM programs (American Council on Education, 2014).

Approximately 5 million post-9/11 service members are expected to transfer out of the military by 2020, many of whom will be covered by the Veterans Educational Assistance Act of 2008 (P.L. 110-252, H.R. 2642). The law specifies that veterans who will have served at least 36 months since September 11, 2001, may receive 100 percent payment for up to 36 months for either all resident tuition and fees for a public college or university or the lower of the actual tuition and fees or the national maximum per academic year for a private college or university. The percentage payment is proportionally lower for shorter military service. One concern about the program is its maximum length of only 36 months, given the small percentage of students who typically complete a 2-year or 4-year degree on time (see Chapter 2).

---

[10] See http://www.nsf.gov/publications/pub_summ.jsp?ods_key=nsf12564 [April 2015].
[11] See http://mesa.ucop.edu/ [April 2015].

PRICE AND COST OF DEGREES

As noted above, barriers to transferring can extend the time to earn a bachelor's degree, thus extending its direct cost to students. Whether related to transfer or not, the cost of earning a degree is another issue that affects all undergraduate students, including those aspiring to earn a STEM degree.

In this section, we describe what is known and not known about the price students pay for STEM degrees relative to other degrees; the cost of delivering STEM degrees relative to non-STEM degrees; and the extent to which differences in tuition price between STEM and non-STEM fields affect demand for STEM majors or completion of STEM degrees. When data are available, these topics are examined for underrepresented minority groups in STEM fields (Hispanics, blacks, Native Americans), and women and for students from low-income families. We begin with a brief discussion of student debt, which is closely related to issues of college costs and prices and provides a context for the rest of the section.

Student Loans and Debt

The percentage of students receiving Pell Grants, which are need-based grants to low-income undergraduate and certain students working on higher degrees, has increased in recent years: it was 25 percent in 2007–2008 and 36 percent in 2012–2013 (The College Board, 2014).

Unfortunately, the students most in need of financial aid may not get it, partly because they do not complete the documentation needed to receive aid in a timely fashion. Recent research in the Chicago Public Schools showed a strong positive correlation between completing the Federal Application for Free Student Aid (FAFSA) and college attendance (Feeney and Heroff, 2013). Using the Illinois Monetary Award Program as a case study, the study found that three student characteristics were related to timely completion of FAFSA: (1) having a slightly higher-than-expected family contribution; (2) having at least one parent who attended college; and (3) demonstrating higher academic performance in high school.

Students who need financial aid can lose their eligibility because they exceed the maximum amount of time to degree completion allowed by the program or because they do not take enough credits to qualify for the program. Students in remedial courses (for which college credits are not awarded) or who transfer are more likely to become ineligible. Student debt has increased significantly over 10 years, even with more students receiving financial aid (The College Board, 2014). For graduates of public 4-year colleges, average debt increased by 12 percent between 2001–2002 and 2006–2007 (from $10,600 to $11,900 in 2012 dollars) and by an additional 20 percent over the next 5 years, to 2011–2012 (from $11,900 to $14,300 in 2012 dollars). For graduates of private nonprofit 4-year colleges, average debt increased by 23 percent between 2001–2002 and 2006–2007 (from $15,400 to $19,000) and by an additional 3 percent over the next 5 years, to 2011–2012 (from $19,000 to $19,500). About 60 percent of students who earned bachelor's degrees in 2011–2012 from public and private nonprofit institutions graduated with debt. These students borrowed an average of $26,500 (The College Board, 2014).

Students with the greatest financial needs have lower rates of degree completion than other students. This is not necessarily a causal relationship, as factors other than finances (i.e., anxiety over financial aid, precollege preparation) may contribute to completion or noncompletion. More research is needed to clarify the relationship among financial need, degree completion, and other related factors.

Price to the Student of a Degree

Determining and evaluating the price to the student of a college education are complex due to variations in costs across institution types, across states, and sometimes even within institutions (due to differential tuition, discussed below). The net price to the student of a college education depends in part on institutional fees, room and board costs, time to degree, number of credits required, and financial aid received.

A significant contributor to rising debt is the increasing price of college since 1999, both absolutely and relative to other costs (Kirshstein, 2013a; see Figure 5-2). The price of a bachelor's degree has increased faster than the rate of inflation, in part due to rising tuition rates. For the 5-year period between 2008–2009 and 2013–2014, the average annual tuition and fees at public 4-year institutions increased 19 percent; for private nonprofit 4-year institutions, the increase was 14 percent. Accompanying the increase in tuition and fees has been an increase in borrowing—a 9 percent growth in total annual educational borrowing between 2007–2008 and 2012–2013.

Policy makers' attention has increasingly focused on the cost of higher education and ways to reduce it. The most ambitious approach is in Tennessee, which enacted the "Tennessee Promise" in 2014, a statewide initiative to provide free community college scholarships and mentoring to all high school graduates. The initiative covers tuition costs and fees not met by existing scholarship programs.[12] More broadly, President Obama has issued a call to make community college free across the nation.[13]

---

[12]For more information on Tennessee Promise, see http://driveto55.org/initiatives/tennessee-promise/ [June 2015].

[13]For details on the President's proposal, see https://www.whitehouse.gov/blog/2015/01/08/president-proposes-make-community-college-free-responsible-students-2-years [June 2015].

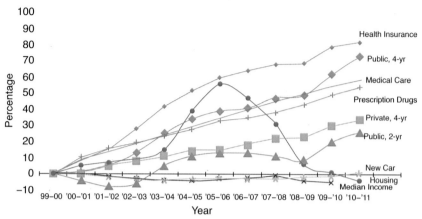

FIGURE 5-2 Percentage change in tuition costs relative to other costs, 1999–2011.
SOURCE: Kirshstein (2013a, p. 2).

The calculated price of a degree varies, depending on such factors as in-state residency (for public institutions), the numbers of credit hours, and range of expenses considered (e.g., tuition and fees, room and board). To facilitate comparing information across available studies, the main focus here is the price of in-state tuition and fees for undergraduates enrolled full time at public institutions (or the price of tuition and fees for full-time undergraduates at private institutions), who have completed or are near completing a bachelor's degree at a public or private institution.

The net price of a degree matters the most for college access and affordability. The net price is the price actually paid by individual students minus the amount of financial assistance received in the form of grants and tax credits. In 2012–2013, undergraduate students as a whole received $185.1 billion in financial aid (The College Board, 2013). Students received 52 percent of this funding in the form of grants, 39 percent as loans (including nonfederal loans), and 9 percent in a combination of tax credits or deductions and Federal Work Study grants.

Price to the Student of a STEM Degree

As shown in Figure 5-3, for students expecting to graduate during the 2007–2008 academic year (the most recent comparable data available), the average net price for a STEM degree ranged from $7,800 for underrepresented minority students at a public 4-year institution to nearly $30,000 for students not from an underrepresented minority group at a private research

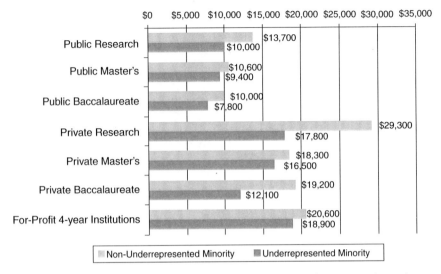

FIGURE 5-3 Net price of an undergraduate STEM degree, by type of institution, for students expected to earn their degree during the 2007–2008 academic year.
SOURCE: Kirshstein (2013b, p. 9).

institution.[14] In the same institution type, the difference in net price paid for physical and natural science and engineering degrees in comparison with social and behavioral science degrees was generally small (Kirshstein, 2013b). Across all institutions, underrepresented minority students pay less than other students, with the largest percentage difference in tuition among students at private research institutions ($17,800 and $29,300, respectively). It is likely that underrepresented minority students pay less than other students both because they typically come from families of greater need (thus, more likely qualifying for tuition assistance) and because they attend institutions with lower price tags, including minority-serving institutions, community colleges, and less-selective 4-year institutions.

The net price for underrepresented minority students who attended public research universities was slightly less than $10,000 (Kirshstein, 2013b). These minority students accounted for 44 percent of STEM degree earners from public research universities. The highest net price for underrepresented minority students was for those students who attended private, for-profit institutions, $18,900: this price was more than $1,000

---

[14]The net price of a degree was calculated as the difference between all student expenses—tuition and fees, room and board, books and supplies, transportation, and other education-related expenses—and the sum of the grants received.

higher than the price paid by students who attended private, not-for-profit research universities.

There are no comparable data of the net price for STEM and non-STEM degrees. However, the average price for a STEM degree in the 2007–2008 academic year was still higher than the price of a non-STEM degree 6 years later (The College Board, 2014).

STEM Student Debt

Figure 5-4 shows the proportion of students who graduated in 2007–2008 with a bachelor's degree in a STEM field (other than psychology and the social sciences) with more than $30,000 of debt. About 65 percent of all STEM majors graduated with debt, in comparison with 60 percent for graduates from 4-year institutions in 2011–2012 (The College Board, 2014). The lowest rates of debt are among students at public research and master's institutions, and the highest rates of debt are among students at for-profit 4-year institutions.

As shown in Figure 5-5 (Kirshstein, 2013c), among STEM majors (in fields other than psychology and the social sciences), larger fractions of students from underrepresented minority groups graduated with a debt of more than $30,000 than did students from other groups. The data for stu-

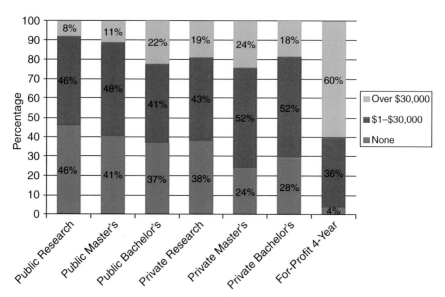

FIGURE 5-4 Percentage of all undergraduate STEM students with various debt levels by type of institution.
SOURCE: Kirshstein (2013c, p. 1).

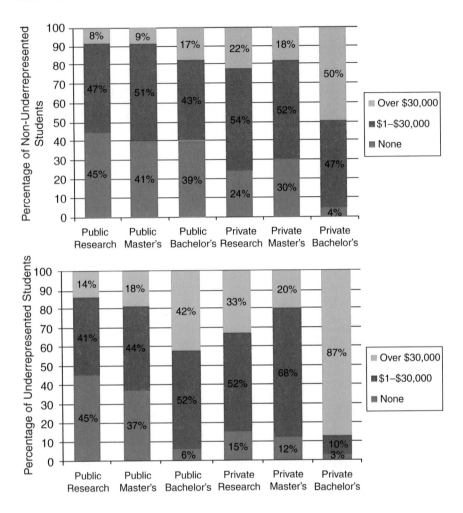

FIGURE 5-5 Undergraduate debt in STEM by minority status.
SOURCE: Kirshstein (2013c, p. 2).

dents in private bachelor's institutions are particularly striking: 87 percent of underrepresented minority students and 50 percent of other students had debt of more than $30,000. In all types of private institutions, nonminority students were more likely than minority students to be debt free. Again, the largest disparity between underrepresented minority students and other students in amount of debt is in public bachelor's institutions: only 6 percent of students from underrepresented minority groups graduate with no debt, in comparison with 39 percent of other students.

Cost to the Institution of a Degree

Total spending of colleges and universities is correlated with the price that students pay and the services they receive. Calculations of total spending typically include education and related spending and general spending (e.g., food services and bookstores). Education and related spending is the common metric used to measure the full "production cost" of education, but it captures only the spending related to services and infrastructure that supports learning, which includes instruction, research, public service, student services (e.g., admissions, registrar services, and student counseling), academic support (e.g., libraries and computing), institutional support (e.g., executive management, and legal and fiscal operations), and associated operation and maintenance. Although other measures of cost have been proposed (see Johnson, 2009), most available data on college costs are based on education and related spending. Thus, the discussion of cost in this section is derived from analyses of education and related spending.

Considering only those categories of costs, the estimated average cost of producing a bachelor's degree at a public 4-year institution in 2009 ranged from about $45,000 to $60,000 in 2009 (Desrochers, 2011). The lower estimate is conservative and includes only the cost of completing 120 credit hours as required by most degree programs. The inclusion of classes taken beyond those required, a situation often experienced by students who transfer institutions, increases costs by 12 percent, to an average of $50,700. The upper estimate includes all of the education and related costs that institutions incur (including credit hours required for a degree, credits that exceed degree requirements, student attrition, and course offerings for nondegree-seeking students). In 2010, the cost per degree declined at most types of institutions as costs were cut in response to reductions in various forms of state tuition subsidies in the context of a sharp economic downturn. However, the cost per degree was still higher than it had been 5 and 10 years earlier. Community colleges consistently decreased costs across the decade (American Institute for Research, 2012).

Another way to analyze the cost of higher education is based on the amount spent per full-time student per year (American Institutes for Research, 2012; see Figure 5-6). For 2-year and 4-year institutions, the costs in 2010 ranged from $9,501 for community colleges (9% more than in 2000) to $12,740 for a public 4-year institution (6% more than in 2000) to $21,126 for a private 4-year institution (12% more than in 2000) and $35,068 for a private research institution (22% more than in 2000). Although funding from state and local appropriations for public institutions is cyclical, the overall trend clearly has been downward. A decade ago, public funds for higher education exceeded tuition revenues from students in public 4-year institutions by $3,000–$5,000 per student, but by 2010,

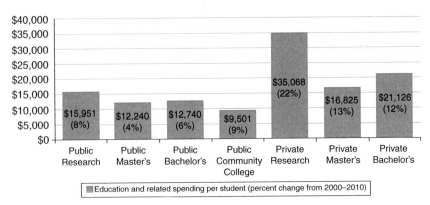

FIGURE 5-6 Annual spending per student in in 2010 for different types of institutions.
SOURCE: Adapted from American Institute for Research (2012, Figure 2).

this difference was about $500 per student. The expectation for the future is that students attending public institutions will continue to pay a larger share of their educational costs.

An analysis by the College Board (2014) found that for the decade between 2000–2001 and 2010–2011, the percentage of institutional revenue from net tuition and fees increased from 21 percent to 34 percent at public 2-year colleges; from 30 percent to 46 percent at public 4-year colleges; and from 88 percent to 94 percent at private nonprofit 4-year colleges (The College Board, 2014). For public institutions, these increases reflect an effort to recover revenue in the face of reduced state and local appropriations.

Along with the downward trend in overall funding from state and local appropriations, there are a growing number of states moving toward performance-based funding for public institutions. This approach uses measures of institutional quality to determine the amount of funding allocated to 2-year and 4-year institutions. Performance-based funding was first implemented by the Tennessee Higher Education Commission in 1978. By 2000, the outcomes-based model for funding was being used in 26 states (Harnish, 2011). Newer versions of performance-based funding are increasingly focused on outcomes deemed important by a state (e.g., graduation rates, average wages of graduates, percentage of students with Pell Grants), and they account for a greater percentage of institutional base funding.

Advocates of performance-based funding argue that the approach advances state goals to improve overall levels of educational attainment and responds to the public's desire to get what it pays for. Some of those who argue against performance-based funding believe it is another effort on the part of state policy makers to cut funding. While the goal of performance-based funding is to increase graduation rates, it has been associated with no changes or negative changes in graduation rates (Tandberg and Hillman, 2013). Others are concerned that financially rewarding completion instead of access will penalize institutions in high poverty areas with the students who are most at risk, while rewarding institutions that serve a student population that is more likely to succeed. Still others believe it will shift the mission of institutions to the point that underprepared students will lose access or that quality will suffer in an effort to move students through to completion.

Beyond the direct cost of producing bachelor's degrees, the high cost of student attrition also affects institutions' costs. Approximately 30 percent of students do not return after their first year (Knapp et al., 2012). Some institutions include this cost in their budgets as part of general or overhead spending or the cost of business. This cost is borne by universities, students, and taxpayers. Data from the U.S. Department of Education's Integrated Postsecondary Education Data System (IPEDS) show that between 2003 and 2008 (the latest years for which data are available), states provided more than $1.4 billion and the federal government provided over $1.5 billion in grants to students who did not return to the institutions at which they were enrolled for a second year. However, the IPEDS data do not show whether the students who did not return for their second year at that institution did not continue at all or continued their education elsewhere. Nor do the IPEDS data show whether such students enrolled at the same or a different institution sometime later.

Cost of a STEM Degree

National data are not available on the cost of a STEM degree. Very few states and institutions collect data on costs at the level of individual disciplines. Estimates of trends from three states (Florida, Indiana, and Ohio) indicate that most STEM degrees (other than psychology and the social sciences) cost more for institutions to produce than non-STEM degrees (Kirshstein, 2013b; see Figure 5-7).[15] Degrees in engineering, engineering technologies/technicians, computer and information sciences, physical sciences, and biological and biomedical sciences cost more than the average cost of a degree across all fields of study, while mathematics and statistics cost less than the average. Social sciences and psychology also cost less than the average. Perhaps in response to these different costs, it is becoming in-

---

[15] These costs were estimated by combining institutional-level data on educational expenditures and degrees from the IPEDS with state-level data on discipline-level credit hour costs from the Four-State Cost Study of 28 disciplines in public baccalaureate-granting institutions that include Florida, Illinois, Ohio, and New York (Conger et al., 2010).

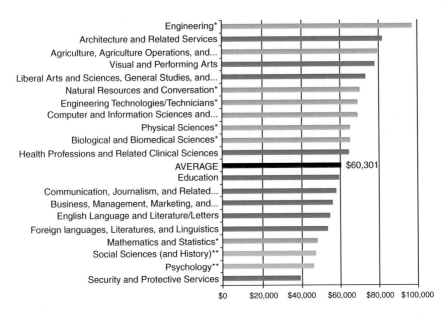

FIGURE 5-7 Estimated cost of providing an undergraduate degree by discipline.
NOTES: * indicates a STEM field; ** indicates a field of social or behavioral science. Institutional-level data on educational expenditures and degrees were combined with state-level data on discipline-level credit hour costs (from Florida, Illinois, and Ohio) to construct various measures of degree production costs. The estimates should not be interpreted as precise costs, but the patterns and trends appear durable.
SOURCE: Kirshstein (2013b, p. 14).

creasingly common for institutions to charge differential tuition for STEM majors, as discussed below.

Differential Tuition and STEM

The convention of charging all undergraduates the same price for full-time study is changing. The enactment of differential pricing practices in which students are charged more tuition for upper- versus lower-division coursework, for example, has grown steadily since the mid-1990s and shows no sign of abating (Cornell Higher Education Research Institute, 2011). In a survey of 165 public research universities, 45 percent reported having differential pricing policies; most of the policies had been implemented in the last decade (Nelson, 2008). A more recent survey reported that 57 percent of public research universities had adopted differential pricing (Reed, 2011). Also, 40 percent of doctoral degree-granting public universities had differential pricing, with the majority assessed according to major (Cornell Higher Education Research Institute, 2011). While some institutions have only recently begun to implement differential tuition, some large public universities, such as the University of Illinois and the University of Michigan, have long charged more for certain coursework or majors (Cornell Higher Education Research Institute, 2011).

Institutions report adopting differential pricing in response to the overall rising costs of program delivery and the need to cover more of their costs with tuition, given the reduced financial support they are receiving from states (Stange, 2013). The relatively high cost of offering STEM degrees is one justification provided for charging more for a STEM degree than for other degrees. As discussed above, although there are limited national data on the cost of a degree at the discipline level, the best available data suggest that institutions are likely to incur more costs to deliver STEM degrees. Engineering, business, and nursing are currently the three degrees most often targeted for differential pricing. A study of 142 large public research universities from 1990 to 2010 showed that 50 of these institutions adopted differential pricing for those degrees during that time (Stange, 2013).

Economists argue that the practice of differential pricing means that students are paying a price that aligns more closely with the actual institutional cost of delivering the degree received, thereby eliminating the subsidization of STEM degrees by students who are majoring in-other fields. It is also argued that differential tuition can better align the price of a degree with a student's ability to pay after graduation. For instance, engineering, science, and business majors tend to earn more and have higher returns on their education investments than students in other majors, such as education and humanities. Thus, students who graduate with the former degrees are said to be in a better position to finance higher tuition fees with loans.

An alternative view, however, is that differential tuition may decrease participation in the fields often targeted for higher prices, such as STEM, especially for low-income students. Given the barrier that the price of a STEM degree can play, especially for students from families with significant financial needs, it is important to understand what is known about how differential tuition practices and policies may contribute to inequalities in STEM degree completion.

Many factors affect students' choice of major, including both nonfinancial factors (e.g., ability, preference, prestige, preparation) and financial considerations (e.g., expected future earnings and ability to pay). Research on the effects of differential pricing is extremely limited in general and especially with respect to how it may interact with other factors to affect students' choice of degree.

Some evidence from a study of students pursuing engineering degrees at two public research universities (George-Jackson et al., 2012) suggests that financial aid initially offsets the higher prices paid for an engineering degree as a result of differential tuition, but that the costs increase over time, particularly for low-income students.

In a controlled survey that compared 50 public research institutions that had adopted differential tuition for engineering, business, and nursing with similar institutions that had not, the choice of major for women and underrepresented minority students was negatively affected by differential tuition policies (Stange, 2013). Such results suggest that differential pricing may deter those groups from majoring in STEM fields. They also indicate that revenues may not be realized or costs shifted as institutions expect (Stange, 2013). However, it is important to note that enrollment in engineering majors decreased in association with differential pricing, while enrollment in nursing programs increased, and enrollment in business programs showed no change.

In a study of student perceptions of differential pricing policies, the policies were viewed as a sign of program quality, but they were also perceived as unfair (Harwell, 2013). Choice of major for the majority of these students was not affected, however.

DEPARTMENTAL POLICIES THAT CAN AFFECT THE COST OF STEM DEGREES

A department's policies, including sequencing of courses, degree requirements and prerequisites outside of the major, grading policies, and remedial course work, can affect the price that a student pays for a STEM degree.

Rigid course sequencing of many majors can increase the cost that a student pays for a STEM degree. The classes that students need to take are not always available, or there may not be enough space for them during the semester that they need to take the courses to meet the required sequence and graduate on time. Students who transfer from another institution may enter with enough credits to be considered a second-year student but may not be able to take 200-level courses until completing all of the required course work that was not taken at the institution from which they transferred. For STEM majors, the high proportion of required credits compared with the proportion of elective credits narrows students' flexibility in meeting requirements. If students cannot take the courses they need in a timely manner, it could result in longer time at college and greater numbers of credits than actually required. Some departments have taken steps to loosen course sequencing and reduce the number of major credit hours required: see Box 5-4 for a description of the steps taken by one small liberal arts college.

Community colleges are also experimenting with ways to reduce the cost of a degree and time to degree, both at the departmental level and

---

BOX 5-4
Departmental Changes at Colby College

When staff and leaders at Colby College recognized that course sequencing and the high number of credit hours required in some majors caused a bottleneck for many students, they undertook a series of steps to address the problem. First, after much debate in committees and then by the faculty as a whole, it was agreed that no major in the college can require more than 50 percent of the credits required for graduation, including prerequisite or co-requisite courses. For some STEM departments, only minor adjustments were required. In others, faculty had to rethink their programs in fundamental ways to decide how to either eliminate requirements for completion of specific courses or to combine expected outcomes so that students could complete the requirement through alternative pathways.

Second, the college changed its graduation requirements, which had included three elements: requirements for the major, other all-college requirements (such as English composition and a course in mathematics), and at least two courses in fields outside a student's major. In the new system, in addition to limiting the number of credits required for a major, students had to be exposed to various areas of knowledge, such as natural science, fine arts, diversity, and quantitative reasoning. At the same time, faculty members modified some of their courses to satisfy the new requirements. For example, a course in logic in the Department of Philosophy might satisfy the requirement for quantitative reasoning if appropriately constructed. Similarly, a course in biology that focused a significant component on discussions of gender might be considered as addressing diversity.

Third, individual departments reorganized some of the requirements for majors to allow more flexibility. For example, the Department of Biology revised a set of rigid requirements for biology to a program that required (1) the introductory course sequence, (2) one course each with lab that focused on three broad areas of biology (molecular/cellular, organismal, and population), and (3) an additional upper-level course. This change in emphasis allowed students to pursue a broader array of pathways through the major and provided them with opportunities to both focus on areas of biology that were of special interest to them while also acquiring the breadth of knowledge that would prepare them for more advanced study, teaching careers at the K-12 level, or other biology-related positions requiring a bachelor's degree. Such an approach can also engender faculty discussion about broad issues—such as "what do WE as a department want our students to know and be able to do once they complete our major?"—rather than focusing on outcomes primarily at the level of individual courses (see, e.g., National Research Council, 2003).

---

more generally, while preparing students for high-demand careers. For example, the 10 community colleges that comprise the U.S. Department of Labor funded National STEM Consortium (NSC)[16] are working together to develop a set of one-year certificate programs that meet regional

---

[16]For details, see http://www.nationalstem.org/ [June 2015].

industry needs. The NSC has developed a two-part STEM bridge program to quickly and efficiently ensure that students are prepared for the rigors of the 30-credit certificate programs. In addition, the certificates offered through the NSC can be earned through a compressed schedule, which allows students to complete the program quickly and be (re)deployed to an employer sooner than with traditional college programs. After completing their certificate, students can work with NSC industry partners to connect with potential employers in their region.

SUMMARY

Many institutional, state, and national undergraduate education policies are not well situated to support students as they progress through the complex and varied pathways to a STEM degree, including transferring among institutions. Research to date suggests that changing policies to increase the transfer of community college course credits could have significant positive effects on student retention and degree completion. Strong articulation agreements among 2-year and 4-year institutions, including common general education requirements, common introductory courses with common numbering, and easily available access to information on course equivalencies across institutions, can improve the percentage of courses transferred and student success.

Regional accrediting agencies, state policy makers, and professional societies can take leadership on requiring institutions to track and share data on their acceptance of transfer credits and also develop and share other metrics of student success, post-transfer. These agencies, policy makers, and societies should reconsider the current model of course-by-course articulation based on content and overseen by individual partnerships of 2-year and 4-year colleges. Changes to support smoother transfer experiences can also be made in institutions, where department leaders (e.g., chairs, deans) are critical actors. Departments often have the latitude to implement policies that can smooth the transition process for students, such as policies that simplify the credit transfer process and provide students with mentoring and other supports needed to successfully transfer.

Undergraduate credentials and degrees in many STEM fields are widely believed to cost more to deliver than degrees in other fields, and there is some evidence to support this belief. In response, some institutions have instituted differential tuition policies. One study shows that differential pricing was associated with a decline in enrollment rates in engineering and business majors over a 3-year period, while it was associated with an increase in enrollment in nursing (Stange, 2013). There are concerns that such policies will have a chilling effect, especially on attracting students from underrepresented groups who already have high levels of borrowing related to unmet financial need. Increases in the price that students pay for a STEM degree, especially in public institutions, appear to be related to decreased support for higher education from state and local sources.

REFERENCES

Alfonso, M. (2006). The impact of community college attendance on baccalaureate attainment. *Research in Higher Education, 47,* 873–903.

American Council on Education (2014). *Higher Ed Spotlight: Undergraduate Student Veterans.* Available: http://combat2career.com/blog/ace-undergraduate-student-veteran-infographic-november-2014/ [February, 2015].

American Institute for Research. (2012). *College Spending in a Turbulent Decade: Findings from the Delta Cost Project: A Delta Data Update 2000–2010.* Available: http://www.deltacostproject.org/sites/default/files/products/Delta-Cost-College-Spending-In-A-Turbulent-Decade.pdf [April 2015].

Anderson, G.M., Sun, J.C., and Alfonso, M. (2006). Effectiveness of statewide articulation agreements on the probability of transfer: A preliminary policy analysis. *Review of Higher Education, 29*(3), 261–291. Available: http://muse.jhu.edu/login?auth=0&type=summary&url=/journals/review_of_higher_education/v029/29.3anderson.html [April 2015].

Austin, A.E. (2011). *Promoting Evidence-Based Change in Undergraduate Science Education.* Paper prepared for the Committee on Barriers and Opportunities in Completing 2-Year and 4-Year STEM Degrees. Available: http://sites.nationalacademies.org/cs/groups/dbassesite/documents/webpage/dbasse_072578.pdf [April 2015].

Bozick, R., and Lauff, E. (2007). *Education Longitudinal Study of 2002 (ELS:2002): A First Look at the Initial Postsecondary Experiences of the Sophomore Class of 2002* (NCES 2008-308). Washington, DC: National Center for Education Statistics, Institute of Education Sciences, U.S. Department of Education.

The College Board. (2011). *Improving Student Transfer from Community Colleges to Four-Year Institutions: The Perspective of Leaders from Baccalaureate-Granting Institutions.* Available: http://www.jkcf.org/assets/1/7/Perspective_of_Leaders_of_Four-Year_institutions.pdf [April 2015].

The College Board. (2013). *Trends in Student Aide 2013: 30 Years 1983–2013.* Available: http://trends.collegeboard.org/sites/default/files/student-aid-2013-full-report.pdf [October 2015].

The College Board. (2014). *Trends in College Pricing 2013.* Available: https://trends.collegeboard.org/sites/default/files/college-pricing-2013-full-report.pdf [April 2015].

Complete College America. (2011). *Time Is the Enemy. The Surprising Truth About Why Today's College Students Aren't Graduating . . . and What Needs to Change.* Available: http://completecollege.org/docs/Time_Is_the_Enemy.pdf [April 2015].

Conger, S.B., Bell, A., and Stanley, J. (2010). *Four-State Cost Study.* Boulder, CO: State Higher Education Executive Officers. Available: http://files.eric.ed.gov/fulltext/ED540266.pdf [April 2016].

Cornell Higher Education Research Institute. (2011). *2011 Survey of Differential Tuition at Public Higher Education Institutions.* Available: https://www.ilr.cornell.edu/sites/ilr.cornell.edu/files/2011%20Survey%20of%20Differential%20Tuition%20at%20Public%20Higher%20Education%20Institutions.pdf [April 2015].

Council of Regional Accrediting Commissions. (2015). *Statement of the Council of Regional Accrediting Commissions (C-RAC) Framework for Competency-Based Education.* Available: https://www.insidehighered.com/sites/default/server_files/files/C-RAC%20CBE%20Statement%20Press%20Release%206_2.pdf [April 2016].

Desrochers, D.M. (2011). *Estimating the Production Cost of Degrees by Level and Discipline Using National Data: An Exploration of Methodologies.* Unpublished manuscript. American Institutes for Research, Washington, DC.

Doyle, W. (2006). Community college transfers and college graduation: Whose choices matter most? *Change, 38*(3), 56–58.

Feeney, M., and Heroff, J. (2013). Barriers to need-based financial aid: Predictors of timely FAFSA completion among low-income students. *Journal of Student Financial Aid, 43*(2), 65–85. Available: http://publications.nasfaa.org/cgi/viewcontent.cgi?article=1204&context=jsfa [April 2015].

George-Jackson, C.E., Rincon, B., and Martinez, M.G. (2012). Low-income students in engineering: Considering financial aid and differential tuition. *Journal of Student Financial Aid, 42*(2), 4–24.

Goldhaber, D., Gross, B., and DeBurgomaster, S. (2008). *Community Colleges and Higher Education: How Do State Transfer and Articulation Policies Impact Student Pathways?* CRPE Working Paper #2008-4. Seattle: Center on Reinventing Public Education, University of Washington. Available: http://www.crpe.org/sites/default/files/wp_crpe4_cc_may08_0.pdf [April 2015].

Goldrick-Rab, S. (2006). Following their every move: How social class shapes postsecondary pathways. *Sociology of Education, 79*(1), 61–79.

Gross, B., and Goldhaber, D. (2009). *Community College Transfer and Articulation Policies: Looking Beneath the Surface.* CRPE Working Paper #2009_1. Available: http://files.eric.ed.gov/fulltext/ED504665.pdf [October 2015].

Harnish, T.L. (2011). *Performance-Based Funding: A Re-emerging Strategy in Public Higher Education Financing.* A Higher Education Policy Brief. Washington, DC: American Association of State Colleges and Universities. Available: http://www.aascu.org/uploadedFiles/AASCU/Content/Root/PolicyAndAdvocacy/PolicyPublications/Performance_Funding_AASCU_June2011.pdf [April 2015].

Harwell, E. (2013). *Students' Perceptions of Differential Tuition Based on Academic Program and the Impact on Major Choice.* IDEALS. Available: http://hdl.handle.net/2142/45516 [April 2015].

Hills, J.R. (1965). Transfer shock: The academic performance of the junior college transfer. The *Journal of Experimental Education, 33*(3), 201–215.

Holahan, C.K., Green, J.L., and Kelley, H.P. (1983). A 6-year longitudinal analysis of transfer student performance and retention. *Journal of College Student Personnel, 24*, 305–310.

Hossler, D., Shapiro, D., Dundar, A., Ziskin, M., Chen, J., Zerquera, D., and Torres, V. (2012). *Transfer and Mobility: A National View of Pre-Degree Student Movement in Postsecondary Institutions.* Herndon, VA: National Student Clearinghouse Research Center. Available: http://pas.indiana.edu/pdf/Transfer%20&%20Mobility.pdf [April 2015].

Johnson, N. (2009). *What Does a College Degree Cost?: Comparing Approaches to Measuring "Cost per Degree."* Delta Cost Project issue brief. American Institute for Research. Available: http://www.deltacostproject.org/products/what-does-college-degree-cost-comparing-approaches-measuring-%E2%80%9Ccost-degree%E2%80%9D [April 2015].

Kirshstein, R. (2013a). Rising tuition and diminishing state funding: An overview. *Journal of Collective Bargaining in the Academy, 0*(14). Available: http://thekeep.eiu.edu/jcba/vol0/iss8/14 [April 2015].

Kirshstein, R. (2013b). *The Price and Cost of a STEM Degree.* Presentation at the second meeting of the Committee on Barriers and Opportunities in Completing 2-Year and 4-Year STEM Degrees, National Academy of Sciences, Washington, DC.

Kirshstein. R. (2013c). *How Much Debt Do Science and Bachelor's Degree Recipients Accrue?* Delta Cost Project, American Institutes for Research, Washington, DC. Available: http://www.deltacostproject.org/sites/default/files/products/STEM%20Debt%20for%20Science%20Bachelors.pdf [June, 2015].

Kisker, C.B., Wagoner, R.L., and Cohen, A.M. (2011). *Implementing Statewide Transfer and Articulation Reform: An Analysis of Transfer Associate Degrees in Four States.* Center for the Study of Community Colleges. Available: http://www.cscconline.org/files/8613/0255/2073/Implementing_Statewide_Transfer_and_Articulation_Reform1.pdf [April 2015].

Knapp, L.G., Kelly-Reid, J.E., and Ginder, S.A., (2012). *Enrollment in Postsecondary Institutions, Fall 2010: Financial Statistics, Fiscal Year 2010 and Graduation Rates, Selected Cohorts 2002–07.* Washington, DC: U.S. Department of Education.

Laanan, F.S. (1996). Making the transition: Understanding the adjustment process of community college transfer students. *Community College Review, 23*(4), 69–84.

Laanan, F.S. (1998). *Beyond Transfer Shock: A Study of Students' College Experiences and Adjustment Processes at UCLA.* Los Angeles: University of California.

Laanan, F.S. (Ed.). (2001). *Transfer Students, Trends and Issues: New Directions for Community Colleges, No. 114.* New York: Wiley.

Melguizo, T., Kienzl, G., and Alfonso, M. (2011). Comparing the educational attainment of community college transfer students and four-year rising juniors using propensity score matching methods. *Journal of Higher Education, 82,* 265–291.

Monaghan, D.B., and Attewell, P. (2014). The community college route to the bachelor's degree. *Educational Evaluation and Policy Analysis,* Available: http://epa.sagepub.com/content/early/2014/02/28/0162373714521865 [April 2015].

Mullin, C.M. (2012). *An Indespensable Part of the Community College Mission.* Policy Brief 2012-03PBL. Available: http://www.aacc.nche.edu/Publications/Briefs/Documents/AACC_Transfer_to_LUMINA.pdf [April 2015].

Mustafa, S., Glenn, D., and Compton, P. (2010). *Transfers in the University System of Ohio: State Initiatives and Outcomes 2002–2009.* Columbus: Ohio Board of Regents. Available: https://www.ohiohighered.org/files/uploads/transfer/research/Transfer_Report_071811_Update.pdf [April 2016].

National Center for Education Statistics. (2013). *Digest of Education Statistics 2013.* Washington, DC: U.S. Department of Education.

The National Center for Public Policy and Higher Education. (2011). *Affordability and Transfer: Critical to Increasing Baccalaureate Degree Completion.* Available: http://www.highereducation.org/reports/pa_at/ [April 2015].

National Research Council. (2003a). BIO2010: *Transforming Undergraduate Education for Future Research Biologists.* Committee on Undergraduate Biology Education to Prepare Research Scientists for the 21st Century. Board on Life Sciences, Division on Earth and Life Studies. Washington, DC: The National Academies Press.

Nelson, G. (2008). *Differential Tuition by Undergraduate Major: Its Use, Amount, and Impact at Public Research Universities.* Lincoln: University of Nebraska. Available: http://digitalcommons.unl.edu/cgi/viewcontent.cgi?article=1004&context=cehsedaddiss [April 2015].

Parker, R., Rittling, M., Rowlett, I., and Jenkins, D. (2014). *Strengthening Transfer Pathways: Tips and Tools for Faculty Engagement.* Presented at the American Association of Community Colleges Annual Convention, April 6, Washington, DC. Available: http://ccrc.tc.columbia.edu/presentation/transfer-pathways-engaging-faculty-aacc.html [April 2015].

Reed, L. (2011). UNL tuition may vary by majors. *Omaha World-Herald,* April 27.

Reynolds, C.L. (2012). Where to attend? Estimating the effects of beginning college at a two-year institution. *Economics of Education Review, 31,* 345–362.

Reynolds, C.L., and DesJardins, S.L. (2009). The use of matching methods in higher education research: Answering whether attendance at a 2-year institution results in differences in educational attainment. In J.C. Smith (Ed.), *Higher Education: Handbook of Theory and Research* (pp. 47–57). New York: Springer.

Roksa, J. (2009). Building bridges for student success: Are transfer policies effective? *Teachers College Record, 111*(10), 2444–2478.

Roksa, J., and Keith, B. (2008). Credits, time, and attainment: Articulation policies and success after transfer. *Educational Evaluation and Policy Analysis, 30*(3), 236–254.

Scott, T.P., Thigpin, S.S., and Bentz, A.O. (in press). Transfer learning community (TLC): Overcoming transfer shock and increasing retention of mathematics and science majors. *Journal of College Student Retention: Research, Theory & Practice.*

Shaffer, C.D., Alvarez, C., Bailey, C., Barnard, D., Bhalla, S., Chandrasekaran, C., Chandrasekaran, V., Chung, H.M., Dorer, D.R., Du, C., Eckdahl, T.T., Poet, J.L., Frohlich, D., Goodman, A.L., Gosser, Y., Hauser, C., Hoopes, L.L., Johnson, D., Jones, C.J., Kaehler, M., Kokan, N., Kopp, O.R., Kuleck, G.A., McNeil, G., Moss, R., Myka, J.L., Nagengast, A., Morris, R., Overvoorde, P.J., Shoop, E., Parrish, S., Reed, K., Regisford, E.G., Revie, D., Rosenwald, A.G., Saville, K., Schroeder, S., Shaw, M., Skuse, G., Smith, C., Smith, M., Spana, E.P., Spratt, M., Stamm, J., Thompson, J.S., Wawersik, M., Wilson, B.A., Youngblom, J., Leung, W., Buhler, J., Mardis, E.R., Lopatto, D., and Elgin, S.C. (2010). The genomics education partnership: Successful integration of research into laboratory classes at a diverse group of undergraduate institutions. *CBE Life Sciences Education, 9*(1), 55–69.

Smith, P.P. (2010). *You Can't Get There from Here: Five Ways to Clear Roadblocks for College Transfer Students.* Education Outlook. Washington, DC: American Enterprise Institute for Public Policy Research. Available: http://www.aei.org/article/education/higher-education/you-cant-get-there-from-here/ [April 2015].

Stephan, J.L., Rosenbaum, J.E., and Person, A.E. (2009). Stratification in college entry and completion. *Social Science Research, 38*, 572–593.

Strange, K. (2013). *Differential Pricing in Undergraduate Education: Effects on Degree Production by Field.* NBER Working Paper 19183. Cambridge, MA: National Bureau of Economic Research. Available: http://nber.org/papers/w19183 [July 2015].

Tandberg, D.A., and Hillman N.W. (2013). *State Performance Funding for Higher Education: Silver Bullet or Red Herring?* (Policy Brief). Madison, WI: University of Wisconsin–Madison, Wisconsin Center for the Advancement of Postsecondary Education.

Townsend, B.K. (1995). Community college transfer students: A case study of survival. *The Review of Higher Education, 18*(2), 175–193.

Wei, C.A., and Woodin, T. (2011). Undergraduate research experiences in biology: Alternatives to the apprenticeship model. *CBE Life Science Education, 10*(2), 123–131.

Western Interstate Commission for Higher Education, (2014). State Higher Education Policy Database. State Summaries: Articulation, transfer and alignment. Available: http://higheredpolicies.wiche.edu/content/policy/state/summaries/31 [April 2015].

6

Leading and Sustaining Change

Major Messages

- For reforms in science, technology, engineering, and mathematics (STEM) to have systemic and lasting effects, they should be tied to broad-based efforts to improve education.
- Lasting change most often occurs when reform strategies include supporting learning communities and networks aimed at creating leaders and change agents who can scale up and sustain changes; developing intensive national and regional ongoing programs; building faculty and academic leader capacity to use data to create and improve reform efforts; and creating intermediary organizations or supporting a coordinating entity.
- To understand better the effect of systemic reform efforts, research will need to focus on reform strategies that are conceptualized broader than just instructional reform and that examine the interlocking qualities of student success (including preparation, advising, supplemental instruction, pedagogy, faculty culture, and articulation between 2-year and 4-year institutions).

The complex pathways identified in Chapter 2 require many different interventions to support student STEM learning at all levels and among different groups—by departments, institutions, business and industry partners, as well as by accreditors and state legislators. As described in Chapter 3, institutions, departments, and professional organizations can take steps to improve the academic culture and instructional practices that students encounter. Chapter 4 details the policies that contribute to or inhibit students' pathways to STEM degrees. In Chapter 5, we also review the cost and price factors that affect students' progress. The many factors involved in STEM education argue for a systems approach to change, at the institutional level for issues related to articulation, at the federal level in terms of funding support, and at the disciplinary level for issues related to rewards and values among faculty. This chapter reviews research on a systems approach to change in higher education and forms the basis for the committee's conclusions and recommendations.

STEM REFORM EFFORTS TO DATE

The empirical research summarized below illustrates that almost all research related to improving STEM education in 2-year and 4-year institutions has had a narrow focus on faculty pedagogy rather than a systems-level approach (Austin, 2011; Fairweather, 2008; Henderson et al., 2011), and there has been very little research on other issues addressed in this report, such as differential tuition or articulation agreements. Overall, STEM reform has been very narrowly considered, primarily focused on in-classroom innovation and the teaching-learning process, which also relates to the narrow way it has been studied as instructional reform (Fairweather, 2008). Drawing on the work by Seymour and Hewitt (1997), much of the research agenda outlined in reports by the National Research Council (see, e.g., 2003a, 2003b, 2012) or sponsored by the National Science Foundation (NSF) has focused on reforming STEM instruction in the college classroom. Fairweather (2008) and others note improved classroom instruction addresses only part of the laundry list of problems in STEM education, including the STEM "pipeline," partnerships with business and industry to improve success in the professions, student advising, and other areas that have largely been ignored when considering change (see Anderson et al., 2011).

Research focused on why STEM reform efforts (particularly curricular and pedagogical innovation) have been so slow to show any effects has identified several barriers to scaling up known positive teaching approaches and curricular alterations. One of the most significant barriers to reform is that most efforts have been focused on individual faculty diffusion of practice, ignoring the context and ecology in which faculty work, as well as other factors that affect student success (Austin, 2011; Fairweather, 2008; Henderson et al., 2011). The NSF and other funders have largely supported individual faculty researchers to conduct curricular and pedagogical reform projects to provide evidence of efficacy. By disseminating the results of the research in a report or workshop, it was assumed other faculty would adopt the practices that support student success (Beach et al., 2012). Yet, after 30 years of funding individual faculty and disseminating results in research journals and at conferences, there has been no systematic adoption of the practices developed through these funded projects (Austin, 2011; Fairweather, 2008), and there are no nationally representative data available to track instructional practices at all 2-year and 4-year institutions.

SYSTEM-LEVEL APPROACHES TO CHANGE IN STEM EDUCATION

In this section of the report, we summarize the limited empirical research about system-level change in STEM to support student success. However, because the research focus in STEM has been quite narrow and does not touch on many of the issues identified in this report, we also draw on research outside of STEM about systemic change.

As detailed in Chapter 3, instructional reforms are typically carried out by individual faculty and at the department level. Only recently have a small number of universities begun to engage in cross-institutional efforts, many of which have been encouraged by other large-scale groups like accreditors, national higher education associations, or disciplinary societies, which can help support sustained change.

Also recently, some efforts have begun to work across larger institutional units, such as across departments or departments working with disciplinary societies. In addition, higher education associations, such as the Association of Public and Land-grant Universities (APLU), the Association of American Universities (AAU), and the Association of American Colleges and Universities (AAC&U), have initiated efforts to improve undergraduate STEM education across their member universities and colleges.

One such effort that worked with disciplinary leaders to transform departments nationally is the Strategic Programs for Innovations in Undergraduate Physics (SPIN-UP) project of the American Physical Society. The disciplinary leaders conducted case studies of departments that had been successful in supporting and graduating students and whose enrollments had grown in recent years.

The researchers identified characteristics of model departments, and these characteristics were broadly shared across the country. Those characteristics included advising, opportunities for undergraduate research, revised courses, and, especially, introductory courses, faculty culture, and the socialization and preparation of students (Hilborn et al., 2003). This approach helped shape a transformation of physics departments nationally, which in turn increased enrollments and student success over time (Hilborn,

2012). The SPIN-UP example also demonstrates that earlier efforts aimed only at teaching ignore other factors critical to student success. Through the American Society for Engineering Education (ASEE), engineering has a long history of periodically examining the state of education and proposing system-level changes, dating back to the Mann Report (American Society for Engineering Education, 1918) and including the Grinter and Green reports (American Society for Engineering Education, 1955, 1994) and *Creating a Culture for Scholarly and Systematic Innovation in Engineering Education* (American Society for Engineering Education, 2012). Another effort in engineering resulted from pressure by the disciplinary accreditor—Accreditation Board for Engineering and Technology (ABET)—which led to national curricular and pedagogical changes (Lattuca et al., 2006). Working with accreditation helped to scale up the changes. The adoption of engineering education outcomes (Engineering Accreditation Commission of the Accreditation Board for Engineering and Technology, 1997) by engineering schools has been considered one of five major shifts in engineering education in the past 100 years (Froyd et al., 2012). Also, NSF funded the Center for the Advancement of Engineering Education project that conducted research on engineering student pathways, engineering educator teaching practices, and methods to build capacity in the field to conduct engineering education research. Both findings and tools developed from this research have been used to improve engineering teaching at institutions across the country (Atman et al., 2012).

AAC&U's Keck/PKAL STEM Education Effectiveness Framework Project[1] has developed an institutional change framework to help campus leaders translate national recommendations for improving teaching and learning in STEM into scalable and sustainable actions. The framework also addresses other supports that have been recommended, such as advising, co-curricular programs, and transfer policies. Participating campuses in California contributed to the development of an institutional readiness audit and a rubric with benchmarking tools that colleges and universities can use to measure their effectiveness in promoting more learner-centered campus cultures in STEM. Results from the project demonstrated that campuses that used the framework were able to make more progress on their change efforts (Kezar et al., in press). These tools are intended to guide campuses through program, departmental, and, eventually, institutional transformation. The project pays specific attention to program and institutional data that can be used to evaluate student achievement, experiences, and progress (e.g., rates of transfer, retention, and completion) with a focus on minority student success. This project developed a framework to take research findings from this and other reports that can be put into action.

---

[1] For more details, see http://www.aacu.org/pkal/educationframework/index.cfm [June 2015].

SPIN-UP, the ABET criteria, and AAC&U's Keck/PKAL have undergone systematic study. Other efforts are currently under way but have not been studied, and we review some of these, which demonstrate the rising understanding that STEM reform needs to work institutionally and across multiple institutions in order to scale up reform. The Bay View Alliance (BVA), the AAU, and the APLU have each created programs to implement and sustain systemic reforms across a number of institutions of higher education, and are connected to other programs.

The BVA is a consortium of research universities carrying out applied research on the leadership of cultural change for increasing the adoption of evidence-based teaching practices.[2] The BVA does not focus directly on teaching methods; instead, it addresses issues related to leadership, motivation, organizational culture, and change management that support and sustain improved teaching practices. The work occurs in research action clusters that conduct research while member universities implement projects. Members of the consortium work together to identify and evaluate more effective ways for university leaders at all levels to inspire and enable improved teaching and learning. Research about the efficacy of the BVA is promising but just beginning.

The AAU Undergraduate STEM Education Initiative[3] seeks to achieve systemic and sustained improvements in STEM learning at its member institutions, which are major public and private research universities. The initiative supports 8 institutions directly, and 41 others focused on improving STEM education as members of the AAU STEM Network. The goals of the initiative include helping institutions assess the quality of STEM teaching on their campuses, share best practices, and create incentives for their departments and faculty members to adopt effective teaching methods. The initiative has developed a framework for systemic change designed to help institutions assess and improve the quality of STEM teaching and learning, particularly during students' first 2 years of college. A demonstration program at a subset of AAU universities is implementing the framework and exploring mechanisms that institutions and departments can use to train, recognize, and reward faculty members who want to improve the quality of their STEM teaching. By 2017, data will begin to be available about the results of the project.

APLU's Science & Mathematics Teacher Imperative (SMTI)[4] works with public universities to increase the number and improve the quality and diversity of science and mathematics teachers they prepare. SMTI has developed an "analytic framework" that allows faculty and administrators

---

[2] For details, see http://bayviewalliance.org/ [February 2015].
[3] For details, see https://www.aau.edu/policy/article.aspx?id=12588 [February 2015].
[4] For details, see http://www.aplu.org/page.aspx?pid=2776 [February 2015].

to analyze policies, processes, and practices that support effective preparation of science and mathematics teachers. An understanding of the factors required for sustained institutional change, including top leadership commitment and faculty ownership, is key to SMTI efforts on campuses.

The AAC&U, AAU, BVA, and APLU have partnered with the American Association for the Advancement of Science and the National Research Council to create the Coalition for Reform of Undergraduate STEM Education. The coalition's goal is to bring about widespread implementation of evidence-based practice in undergraduate STEM education. The coalition will share data and approaches, monitor progress nationally on metrics and models for institutional change, analyze for gaps, encourage action on gaps, and work to attract funding to this agenda. The coalition is also working to build ongoing capacity within the several partner organizations and their respective STEM educational programs to advance the adoption of evidence-based STEM practices on college, university, and community college campuses.

In addition to the primary outcome of using exchange of information to strengthen the ongoing work on members' initiatives, specific outcomes of the Coalition for Reform of Undergraduate STEM Education have included preparation of an initial matrix of relevant national-level activities and a meeting of practitioners and funders, supported by the Research Corporation for Science Advancement and the Sloan Foundation.[5] The meeting explored ways to deepen and scale needed reforms. Moving forward, the Coalition will strive to focus on mapping the space for reform and promoting commitment to the systemic changes needed to achieve widespread implementation of evidence-based practices.

In line with research on organizational reform (Austin, 2011; Fairweather, 2008; Henderson et al., 2011; Kezar, 2011; Manduca, 2008), the efforts of BVA, AAU, AAC&U, and APLU are designed to create change at multiple levels (institution, discipline and department, and program), rather than focusing on individual faculty or even single departments. The design of the reform efforts is in line with research showing how through departments, leaders can reshape entire curriculum and create professional learning communities focused on new pedagogy (Austin, 2011; Fairweather, 2008; Henderson et al., 2011; Kezar, 2011; Manduca, 2008). The reform efforts also align with research findings that illustrate the importance of conceptualizing STEM reform as part of a complex ecology—departments, institutions, disciplines, national organizations, foundations, accreditors, state policy makers, and other groups that can be leveraged for change (Austin, 2011; Kezar, 2011; Zemsky et al., 2005).

---

[5]For a summary of the meeting, see Achieving Systemic Change at https://www.aacu.org/pkal/sourcebook [June 2015].

Both SPIN-UP and engineering reforms through ABET also highlight how changes in teaching should not be seen in isolation and that success for students means examining student support, departmental climate, and other issues. While these two initiatives did not focus as directly on institutions as the site for change, they allude to many issues related to student success that are beyond departmental control and would require, instead, institutional policies (around incentives, for example), practices (e.g., values supported by awards), and leadership to create meaningful changes for student success.

In addition to the inherent flaw in the narrow approach to scaling change by simple dissemination of information about good practice, there are other identified barriers in institutions to STEM reform that have led to systemic reform strategies. Those barriers are related to how institutions relate to each other and how they affect society. For example, a collaborative effort to scale up a developmental mathematics reform movement in Texas, the New Mathways Project,[6] focused on how the various institutions with a stake in higher education in Texas (state governments, colleges, funding organizations) interact with each other. A major finding from this work was that many institutions are optimizing for legitimacy (or to be perceived as authoritative) rather than for quality (Rutschow et al., 2015). Since institutions tend to seek prestige and status and to copy their peers, incentives for change include making teaching and student success a measure of prestige as is being developed in the AAU initiative. The AAU initiative also tries to create groups of peers or networks that will influence each other over in the long run.

Both the National Science Foundation and the Howard Hughes Medical Institute have funded projects intended to catalyze change at the institutional level. These projects show promise for models that could be adopted or adapted at multiple institutions and can lead to large-scale change.

As discussed above, there are many barriers to change. In a meta-analysis of studies of STEM reform, Henderson and colleagues (2011) identified incentives, reward systems, disciplinary values, and institutional support as key barriers. However, none of the studies they reviewed addressed whether strategies to overcome these barriers—such as new recognition and reward systems—had led to positive change. In addition to barriers (see Fairweather, 2008), there are factors that can influence adoption of new teaching techniques, such as faculty workload, faculty rewards, sequence of courses in curricula, leadership, and resources. Fairweather (1996) provides evidence that the reward system systematically continues to devalue teaching for people in tenure-track positions in 4-year and graduate-level

---

[6]This is a collaborative effort among the Dana Center at the University of Texas at Austin and the Texas Association of Community Colleges.

institutions, a major detriment to change. Furthermore, the increased use of part-time faculty to teach introductory STEM courses also inhibits reform as faculty are cycling in and out of classes (Kezar, 2013).

Looking across the available research, four major weaknesses in previous reform efforts have been identified:

1. Focusing too narrowly on individuals rather than the entire system, which leads to small-scale and short-lived changes.
2. Not leveraging multiple levels—individual, department, institution, disciplinary society, business and industry, government and policy—for change.
3. Focusing too narrowly on pedagogical and curricular changes while not also considering other aspects related to student success.
4. Focusing on a single area, such as undergraduate research, rather than looking at the entire system.

OPPORTUNITIES IN STEM REFORM

There has been relatively little research to provide guidance on what factors promote reform, both generally and specifically in STEM. As noted above, studies have been framed so narrowly that a complex array of student success factors has been tried in few situations and rarely studied.

In their meta-analysis of approaches to reform (related to pedagogy and instruction), Henderson and colleagues (2011) identified four categories of approaches to change that suggest directions for moving forward: (1) disseminating curriculum and pedagogy, (2) developing reflective teachers, (3) enacting policy changes, and (4) developing a shared vision. However, STEM education researchers largely write about change only in terms of disseminating curriculum and pedagogy, and this strategy has led to minimal change. While this strategy has poor efficacy, Seymour (2001) points out that one reason that it may persist is that it is often reflected in proposal requirements of funding agencies.

The least used strategy for change found by Henderson and colleagues (2011) (only 8 percent of articles they reviewed) with the most efficacy is developing a shared vision for the change, often through the creation of learning communities, organizational learning processes, and/or culture change (see below). Moreover, most studies of change provide minimal evidence to support whether the change strategy worked. For example, only 21 percent of the articles that reported on implementation of a change strategy were categorized as presenting strong evidence to support claims of the success or failure of the strategy. They conclude (Henderson et al., 2011, p. 1): "[T]he state of change strategies and the study of change strategies are weak, and that research communities that study and enact change are largely isolated from one another."

Effective instructional change strategies have the following characteristics: they are aligned with or seek to change the beliefs of the individuals involved; they involve long-term interventions, lasting at least one semester and often longer; they require understanding a college or university as a complex system, and they design a strategy that is compatible with this system. In the rest of this section, we present findings from other higher education research that documents approaches to creating systems-level and sustained changes focused on learning communities, organizational learning and data-driven decision-making, and faculty development. We also consider issues of institutional support, multilevel leadership, and multifaceted approaches to change. It is important to note that many of these strategies are aimed at cultural change, and in recent years there has been growing awareness that enhancing STEM learning environments requires a change in the values and beliefs of faculty and academic and disciplinary leaders.

Learning Communities

Various studies in higher education support the idea that changing individual belief systems through discussion and deliberation is important to change and for scaling up interventions (Gioia and Thomas, 1996; Kezar, 2001, 2012, 2013). By changing faculty and staff beliefs, changes are deeper and sustained (Kezar, 2012). One way to support changes in beliefs is learning communities and reform networks.

Most research on learning communities has been conducted on K-12 education, but there is a growing research base that faculty learning communities also lead to change in practice and work to scale up changes across departments (Quardokus and Henderson, 2014). Kezar and Gehrke (2015) found that large national STEM reform networks have the potential to spread reforms across thousands of faculty, as well as help them become change agents who can reshape departments and colleges (see also American Association for the Advancement of Science, 2015).

Learning communities reflect the characteristics found in the Henderson and colleagues (2011) meta-analysis: they engage individuals in changing beliefs, over a long time, and help faculty members understand barriers and facilitators for change in their institutions. When designed appropriately, these networks can help spread and sustain change. There have been few efforts to create regional or national learning communities for STEM reform, but research on reforms in higher education outside STEM suggests that networks and learning communities have been among the most significant vehicles for scaling up such changes as service learning or undergraduate research (Kezar, 2011, 2013; Smith et al., 2004). The importance of learning communities is also demonstrated in Kezar's research on the three key qualities that lead to scale in higher education: provide opportunities for sustained deliberation of change; support ongoing networks and communities for change agents; and develop intermediary organizations to provide incentives, support, and rewards (Kezar, 2011). The creation of learning communities develops both networks and opportunities for sustained discussion, two of the critical elements that can lead to large-scale change. Centers for teaching and learning can also offer an institution-based tool/ resource to support these changes.

Organizational Learning and Data-Driven Decision Making

While STEM reform research has until recently ignored organizational and institutional approaches to change, the notion of learning communities can be connected to research about organizational learning (Fulton and Britton, 2011). Learning communities essentially provide opportunities for groups to learn and change together. When this group approach to learning is institutionalized and expanded into the larger organization, it is labeled organizational learning. Organizational learning has been identified in the broader literature on change as one of the most robust strategies for creating change (Kezar, 2001, 2013; Smith, 2015; Sturdy and Grey, 2013). Organizational learning is a key way to address change since the STEM education problems will vary by institution, and no one can know all the individual issues affecting policy or culture (Kezar, 2013). The introduction of "broader impacts" as one of the review criteria for award of NSF research and education funding has led some institutions to construct institution-level infrastructure (capacity) to support principal investigators. Various units within NSF provide examples of efforts that would fall into the category of broader impacts. Efforts to use the examples from NSF to build "local options" that build on institutional and principal investigator assets have been a powerful driver for change in some institutions.

In general, research on change suggests that organizational learning promotes change by helping prompt doubt in people about their beliefs by presenting data and evidence to guide decision making and thinking (Kezar, 2001, 2012; Senge, 1990). Organizational learning also helps create context-based solutions. For initiatives on student success, research has demonstrated that collecting and analyzing data on students by differences in demographics, majors, course-taking patterns, and pathways and using these analyses to develop interventions has helped increase student success (Bauman, 2005; Bensimon and Malcom, 2012). Chapter 2 detailed the complex pathways in STEM education. Institutions also need to collect and review data about their own students in order to develop appropriate interventions. Data-informed decision making is not without challenges as many organizations lack the infrastructure to collect, aggregate, and interpret the needed data. In addition, data-driven decision making is complicated by the need to focus on success of all students, which would require examining system changes from a variety of perspectives. The recent initiatives noted above, such as the AAU and AACU/Keck Framework projects, use data, metrics, and organizational learning to develop appropriate policies.

Faculty Development

Research studies also support the role of robust faculty development efforts to improve STEM education. Policies created at the national, regional, or institutional level are particularly important for changing faculty instructional practices. As an example, the Physics and Astronomy New Faculty Workshops (NFW) Program, sponsored by the American Association of Physics Teachers, American Physical Society, and American Astronomical Society, and supported by NSF, has offered 17 workshops, each lasting 3 or more days. There is strong evidence to suggest that the NFW Program has been very successful at increasing participant knowledge about research-based instructional strategies and motivating participants to try these strategies (Henderson et al., 2012). In a national survey of randomly selected U.S. physics faculty, those who had attended NFW had the largest correlation of 20 personal and situational variables indicating a respondent's knowledge about and use of at least one research-based instructional strategy (Henderson et al., 2012). But the survey results also show that faculty members who attend professional development but then return to their campuses to find an unfavorable environment for change do not continue reforms.

The AAU, AACU, and Bay View Alliance projects all build on this research and incorporate in-depth faculty development programs into their projects. Faculty development efforts from these organizations also seek to foster a more supportive environment or climate within STEM departments. Institutional-level efforts, such as centers for teaching and learning, can also support a more favorable environment for reform.

Institutional Support

Findings from Fairweather (2008) and Austin (2011), among others, suggest that creating change mechanisms like learning communities or offering professional development without addressing the incentive system and values in academia will largely result in only short-term change. This research suggests even if good practices and changed beliefs are spread, they are unlikely to be sustained if the overall culture and structure of the institution does not support changes. Although there are no long-term studies of whether changes supported by learning communities remain over time, the implication from Fairweather (2008) and Austin (2011) is that current practices are divorced from addressing systemic barriers and will not be successful.

Some, but not all, learning communities address institutional barriers. Organizational learning approaches simultaneously try to identify and address both barriers and solutions. But change efforts aimed at addressing systems barriers, such as rewards for teaching that are embedded into the recent AAU Initiative, are likely pivotal to scale change. Leadership on campus is critical to engage to change reward and incentive systems.

Multilevel Leadership

Research on change in colleges and universities demonstrates that systemic change occurs when leaders across multiple parts of the system work in concert toward a solution (Kezar, 2013). The type of changes outlined in this report will likely not be initiated and sustained unless there is leadership capacity at multiple levels. Leaders can shape and change incentives and rewards, and create more robust systems to enhance data-informed decision making. Leaders are critical to altering the culture by reshaping values and what is considered normative.

Department chair leadership programs have been shown as instrumental to other types of STEM changes, such as getting more women and underrepresented minorities into STEM disciplines as faculty (Rosser, 2009). Chairs can help implement support for students at the departmental level and support instructional and curricular changes. Individual campuses are increasingly offering chair training because they recognize that these individuals are critical to policy implementation, but recent studies (McClelland and Holland, 2015; Quardokus and Henderson, 2014) also demonstrate their role in change.

Deans, provosts, presidents, trustees, and regents are needed to examine policies around tuition, articulation, and course credit. Institutional leaders are known to be significant players in implementing changes, but they are not generally brought into STEM reform efforts (Kezar, 2013). Pressure from external players such as accreditors, legislative bodies, government agencies, and business and industry leaders has also been instrumental and can be used as a lever for STEM reform (Eckel and Kezar, 2003).

Disciplinary leadership is needed to examine ways that professional societies can encourage support for improved teaching, new instructional methods, and strategies to increase student success rates. Disciplines set norms about who is considered a scientist or engineer, and many disciplines may remain unwelcoming places for some STEM students. Disciplinary leadership has been studied less than institutional leadership, but even the

STEM research examples above (e.g., SPIN-UP and the ABET criteria) demonstrate the role leaders have played in certain fields.

The creation of a national group, network, or organization that would bring together STEM reform leaders could help in fostering changes in STEM education at a high level. Organizations such as the Council for Undergraduate Research and Campus Compact bring together leaders across multiple parts of the enterprise in support of these practices. Research demonstrates that interaction has furthered the spread of these practices (Hollander and Hartley, 2000). The recent development of the Coalition for Reform of Undergraduate STEM Education will further increase collaboration among various STEM reform efforts and reform advocates.

Capacity for leadership could be provided through disciplinary societies, national organizations, and individual institutions. For an example of the latter, see Box 6-1. Project Kaleidoscope[7] has a Summer Leadership Institute for department chairs, faculty, and other academic leaders to help them in working to create needed changes. A few disciplinary societies (e.g., American Society for Engineering Education, American Physical Society) have also created leadership development initiatives (see Chapter 3). Studies of leadership development demonstrate the value of understanding the local (institutional) context and the value of learning from individuals in other contexts that can provide a broader (disciplinary) perspective and national view on issues. Both kinds of opportunities are likely the most beneficial.

Leadership for change might be encouraged through the development of a prestigious prize or honor created that is given to a campus or department for exceptional leadership to support student success in STEM. A foundation, association, or agency might be encouraged to develop an award similar to the American Mathematical Society's Award for an Exemplary Program or Achievement in a Mathematics Department.[8] Awards have been found to motivate changes among leaders and alter cultural norms (e.g., the Baldridge and Aspen awards). Awards can also draw attention among the multiple levels of leadership in the system and create a sense of focus for change.

Multifaceted Approaches to Change

As noted above, STEM reform change studies have generally not examined multiple factors at the same time—undergraduate research, advising, and instructional reform. As a result, there is no evidence on whether addressing multiple factors leads to greater student success. However, research in higher education on student success initiatives outside of STEM

---

[7]This is a project of the AAU. For details, see https://www.aacu.org/pkal [June 2015].

[8]For details, see http://www.ams.org/news?news_id=2632 [July 2015].

BOX 6-1
Undergraduate STEM Reform at Georgia State University

In his opening remarks at the Vision and Change event held August 28–30, 2013, Mark Becker, president of Georgia State University (GSU), noted that national policy makers believe 60 percent of the populace needs to have a college education if the nation is to maintain its competitive advantage in the new global economy. Currently, less than 30 percent of the population meets this goal, so something needs to change. In the case of GSU, Becker knew that to improve its graduation rates (which had been at 32%), the university had to become more inclusive and more committed to student success, and *everyone* on campus had to bring a sense of urgency to the task at hand.

To this end, GSU spent close to a year developing a 5-year strategic plan, and testing a variety of approaches that could achieve these goals. Throughout the process, campus administrators collected and analyzed data so that documented successes could be scaled up immediately. For example, the use of trained peer tutors and freshmen learning communities showed positive results, so GSU leaders introduced peer tutors into every class that had high failure or withdrawal rates and required all entering freshmen to join a learning community unless they specifically requested to opt out.

GSU also addressed the financial issues that often force students to withdraw, helping undergraduates secure or regain scholarships or providing small grants to encourage students to attend courses on how to manage their time and finances, while improving their study skills. The campus also centralized its advising structure so that students had opportunities to meet with trained advisers who could help keep them on a path toward graduation.

A summer success program for underprepared students then concentrated all of the successful campus initiatives—for example, peer tutoring, study skills development, and learning communities—into one intensive summer experience conducted before students began their freshman year. This program helped prepare new students for the rigors of college-level work, and participants subsequently performed as well as their better-prepared peers. As a result, graduation rates at GSU have improved from 32 percent to 54 percent in just 3 years.

---

demonstrates that a multipronged strategy for addressing student success leads to greater persistence and higher rates of retention and graduation for students (Bean, 2005; Braxton, 2000; Kuh, 2008; Tinto, 2006). Thirty years of research on student retention and success demonstrates that student success is a complex puzzle that requires attention to college preparation and transition, advising, financial aid, faculty-student interactions, faculty's use of high-quality pedagogy, articulation and transfer polices, engagement in high impact practices, and the like (Bean, 2005; Braxton, 2000; Kuh, 2008). While not every approach may be addressed in each reform effort, focusing more broadly across institutional factors (culture, faculty teaching, financial aid, articulation and transfer) and enterprise factors (rewards, disciplinary norms, prestige seeking, competition between institutions) will likely lead to greater student success over time (Tinto, 2006).

SUMMARY

In this chapter, we illustrate the need for reform efforts to address systemic barriers of disciplinary and institutional value systems. Systemic and lasting reform in education, including STEM education, requires an approach that addresses multiple levels of leadership: department, institution, discipline, government, and business and industry.

Strategies for successful undergraduate STEM reform that emerge from the available research include supporting learning communities and networks (disciplinary, national, and regional) that help change faculty belief systems and practices and that are aimed at creating leaders and change agents who can scale up and sustain changes; developing ongoing national, regional, and disciplinary faculty development programs; providing faculty and academic leaders the capacity to use data to create and improve reform efforts; and creating intermediary organizations or supporting a coordinating entity, such as the Coalition for Reform of Undergraduate STEM Education, to focus and support reform.

To better understand the effect of such reform efforts, research is needed on reform strategies that are broader than just instructional reform and that examine the interlocking qualities of student success, which include preparation, advising, supplemental instruction, pedagogy, faculty culture, and articulation between 2-year and 4-year institutions. A shift toward funding and studying reform efforts that focus on multiple levels could yield significant benefits for all who are involved in undergraduate STEM education.

REFERENCES

American Association for the Advancement of Science. (2015). *Vision and Change in Undergraduate Biology Education: Chronicling Change and Inspiring the Future*. Washington, DC: American Association for the Advancement of Science.

American Society for Engineering Education. (1918). *The Mann Report*. Washington, DC: American Society for Engineering Education.

American Society for Engineering Education. (1955). *Evaluation of Engineering Education ("The Grinter Report")*. Washington, DC: American Society for Engineering Education.

American Society for Engineering Education. (1994). *Engineering Education for a Changing World ("The Green Report")*. Washington, DC: American Society for Engineering Education.

American Society for Engineering Education. (2012). *Innovation with Impact: Creating a Culture for Scholarly and Systematic Innovation in Engineering Education*. Washington, DC: American Society for Engineering Education.

Anderson, W.A., Banerjee, U., Drennan, C.L., Elgin, S.C.R, Epstein, I.R., Handelsman, J., Hatfull, G.F., Losick, R., O'Dowd, D.K., Olivera, B.M., Strobel, S.A., Walker, G.C., and Warner., I.M. (2011). Changing the culture of science education at research universities. *Science, 331*, 152–153. Available: http://www.physics.emory.edu/~weeks/journal/anderson-sci11.pdf [April 2015].

Atman, C., Sheppard, S., Turns, J., Adams, R., Yasuhara, K., and Lund, D. (2012) The Center for the Advancement of Engineering Education: Results and resources. *International Journal of Engineering Education, 28*(5), 1–14.

Austin, A.E. (2011). *Promoting Evidence-Based Change in Undergraduate Science Education.* Paper prepared for the Committee on Barriers and Opportunities in Completing 2-Year and 4-Year STEM Degrees. Available: http://sites.nationalacademies.org/cs/groups/dbassesite/documents/webpage/dbasse_072578.pdf [April 2015].

Bauman, G.L. (2005). Promoting organizational learning in higher education to achieve equity in educational outcomes. In A. Kezar (Ed.), *Organizational Learning in Higher Education: New Directions for Higher Education* (Chapter 2). San Francisco: Jossey-Bass.

Beach, A.L, Henderson, C., and Finkelstein, N. (2012). Facilitating change in undergraduate STEM education. *Change: The Magazine of Higher Learning, 44*(6), 52–59.

Bean, J.P. (2005). Nine themes of college student retention. *College Student Retention*, 215–243.

Bensimon, E.M., and Malcom, L.E. (2012). *Confronting Equity Issues on Campus: Implementing the Equity Scorecard in Theory and Practice.* Sterling, VA: Stylus.

Braxton, J.M. (Ed.). (2000). *Reworking the Student Departure Puzzle.* Nashville, TN: Vanderbilt University Press.

Eckel, P. D., and Kezar, A. (2003). *Taking the Reins: Institutional Transformation in Higher Education.* American Council on Education, Westport, CT: Praeger.

Engineering Accreditation Commission of The Accreditation Board for Engineering and Technology. (1997). *Engineering Criteria 2000, Third Edition.* Baltimore, MD: Engineering Accreditation Commission of The Accreditation Board for Engineering and Technology.

Fairweather, J.S. (1996). *Faculty Work and Public Trust: Restoring the Value of Teaching and Public Service in American Academic Life.* Boston: Allyn & Bacon.

Fairweather, J.S. (2008). *Linking Evidence and Promising Practices in Science, Technology, Engineering, and Mathematics (STEM) Undergraduate Education: A Status Report for the National Academies National Research Council Board on Science Education.* Paper prepared for the Workshop on Evidence on Promising Practices in Undergraduate Science, Technology, Engineering, and Mathematics (STEM) Education. Available: http://www.nsf.gov/attachments/117803/public/Xc--Linking_Evidence--Fairweather.pdf [April 2015].

Froyd, J.E., Wankat, P.C., and Smith, K.A. (2012). Five major shifts in 100 years of engineering education. *Proceedings of the IEEE, 100*, 1344–1360.

Fulton, K., and Britton, T. (2011). *STEM Teachers in Professional Learning Communities: From Good Teachers to Great Teaching.* Washington, DC: National Commission on Teaching and America's Future.

Gioia, D., and Thomas, J. (1996). Identity, image, and issue interpretation: Sensemaking during strategic change in academia. *Administrative Science Quarterly, 41*(3), 370–403.

Henderson, C., Beach, A., and Finkelstein, N. (2011). Facilitating change in undergraduate STEM instructional practices: An analytic review of the literature. *Journal of Research in Science Teaching, 48*(8), 952–984.

Henderson, C., Dancy, M., and Niewiadomska-Bugaj, M. (2012) The use of research-based instructional strategies in introductory physics: Where do faculty leave the innovation-decision process? *Physical Review Special Topics: Physics Education Research, 8*(2). Available: http://journals.aps.org/prstper/abstract/10.1103/PhysRevSTPER.8.020104 [April 2015].

Hilborn, R.C., Howes, R.H., and Krane, K.S. (2003). *Strategic Programs for Innovations in Undergraduate Physics: Project Report.* College Park, MD: American Association of Physics Teachers.

Hilborn, R.C. (2012). *Growing Undergraduate Physics Programs: What SPIN-UP Tells Us Works.* Paper presented at the Joint Spring Meeting of the Texas Sections of the AP and AAPT and Zone 13 of the SPS, March 22-24, San Angelo, TX. Available: http://meeting. aps.org/Meeting/TSS12/Session/G1.1 [April 2015].

Hollander, E., and Hartley, M. (2000). Civic renewal in higher education: The state of the movement and the need for a national network. In T. Ehrlich (Ed.), *Civic Responsibility and Higher Education* (pp. 345–366). Westport, CN: American Council on Education and Oryx Press.

Kezar, A. (2001). Understanding and Facilitating Organizational Change in the 21st Century: Recent Research and Conceptualizations. *VASHE-ERIC Higher Education Report, 28*(4). San Francisco: Jossey-Bass.

Kezar, A. (2011). What is the best way to achieve reach of improved practices in education? *Innovative Higher Education, 36*(11), 235–249.

Kezar, A. (2012). Understanding sense-making in transformational change processes from the bottom up. *Studies in Higher Education, 65*, 761–780.

Kezar, A. (2013). *How Colleges Change.* New York: Routledge.

Kezar, A., and Gehrke, S. (2015). *Scaling up Undergraduate STEM Reform: Communities of Transformation–Their Outcomes, Design, and Sustainability.* Los Angeles: Pullias Center.

Kezar, A., Gehrke, S., and Elrod, S. (in press). Implicit theories of change as a barrier to change on college campuses: An examination of STEM reform. Submitted to *The Review of Higher Education.*

Kuh, G.D. (2008). Excerpt from "High-Impact Educational Practices: What They Are, Who Has Access to Them, and Why They Matter." Association for American Colleges & Universities. Available: https://www.aacu.org/leap/hips [April 2015].

Lattuca, L.R., Terenzini, P.T., and Volkwein, J.F. (2006). *Engineering Change: A Study of the Impact of EC2000, Executive Summary.* Baltimore, MD: ABET.

Manduca, C.A. (2008). *Working with the Discipline: Developing a Supportive Environment for Education.* Presented at the National Research Council's Workshop Linking Evidence to Promising Practices in STEM Undergraduate Education, Washington, DC. Available: http://sites.nationalacademies.org/cs/groups/dbassesite/documents/webpage/ dbasse_072635.pdf [April 2015].

McClelland, S.I., and Holland, K.J. (2015). You, me, or her: Leaders' perceptions of responsibility for increasing gender diversity in STEM departments. *Psychology of Women Quarterly, 39*(2), 210–225.

National Research Council. (2003a). *BIO2010: Transforming Undergraduate Education for Future Research Biologists.* Committee on Undergraduate Biology Education to Prepare Research Scientists for the 21st Century. Board on Life Sciences, Division on Earth and Life Studies. Washington, DC: The National Academies Press.

National Research Council. (2003b). *Improving Undergraduate Instruction in Science, Technology, Engineering, and Mathematics: Report of a Workshop.* Steering Committee on Criteria and Benchmarks for Increased Learning from Undergraduate STEM Instruction. R.A. McCray, R. DeHaan, and J.A. Schuck (Eds). Committee on Undergraduate Science Education, Center for Education. Division of Behavioral and Social Sciences and Education. Washington, DC: The National Academies Press.

National Research Council. (2012). *Discipline-Based Education Research: Understanding and Improving Learning in Undergraduate Science and Engineering.* Committee on the Status, Contributions, and Future Directions of Discipline-Based Education Research. S. Singer, N.R. Nielsen, and H.A. Schweingruber (Eds.). Board on Science Education, Division of Behavioral and Social Sciences and Education. Washington, DC: The National Academies Press.

Quardokus, K., and Henderson, C. (2014). Promoting instructional change: Using social network analysis to understand the hidden structure of academic departments. *Higher Education.* Available: http://link.springer.com/article/10.1007%2Fs10734-014-9831-0 [June 2015].

Rosser, S.V. (2009). Creating a new breed of academic leaders from STEM women faculty: NSF's ADVANCE Program." In A. Kezar (Ed.), *Rethinking Leadership in a Complex, Multicultural, and Global Environment* (pp. 117-130). Sterling, VA: Stylus.

Rutschow, E.Z., Diamond, J, and Serna-Wallender, E. (2015). *Laying the Foundations: Early Findings from the New Mathways Project.* New York: MDRC.

Senge, P. (1990). *The Fifth Discipline: The Art and Practice of the Learning Organization.* New York: Doubleday.

Seymour, E. (2001). Tracking the process of change in U.S. undergraduate education in science, mathematics, engineering, and technology. *Science Education, 86,* 79–105.

Seymour, E., and Hewitt, N. (1997). *Talking about Leaving: Why Undergraduates Leave the Sciences.* Boulder, CO: Westview Press.

Smith, D. (2015). *Diversity's Promise for Higher Education: Making it Work; Second Edition.* Baltimore, MD: Johns Hopkins University Press.

Smith, B.L., MacGregor, J., Matthews, R.S., and Gabelnick. F. (2004). *Learning Communities: Reforming Undergraduate Education.* San Francisco: Jossey-Bass.

Sturdy, A., and Grey, C. (2013). Beneath and beyond organizational change management: Exploring alternatives. *Organization, 10*(4), 651–652.

Tinto, V. (2006). Research and practice of student retention: what next? *Journal of College Student Retention: Research, Theory and Practice, 8*(1), 1–19.

Zemsky, R., Wegner, G., and Massy, W.F. (2005). *Remaking the American University: Market-Smart and Mission-Centered.* New Brunswick, NJ: Rutgers University Press

7

Conclusions and Recommendations

Students who enter college to earn a 2-year or 4-year degree in an area of science, technology, engineering, and mathematics (STEM) face many barriers in the multiple pathways to degree completion. The pathways that students are taking to earn STEM degrees are diverse and complex, with multiple entry and exit points and an increased tendency to earn credits from multiple institutions. The barriers students face differentially affect students from underrepresented minority groups and women, as shown by the lower rates of degree completion by black, Hispanic, and female students. The barriers are particularly difficult to overcome for students with limited experience with and knowledge of higher education in general and of STEM fields in particular, such as first-generation students and many of those who are eligible for Pell Grants. The undergraduate student population has undergone significant shifts, and undergraduates who aspire to earn STEM degrees are much different than their counterparts 25 years ago. The percentage of women and students from underrepresented backgrounds who are interested in STEM degrees has been on the rise (National Science Board, 2014). The number of students attending undergraduate institutions who have previous work experience, have taken a semester or more away from college, and have families is also increasing (National Center for Education Statistics, 2013). And as noted throughout this report, students interested in STEM degrees are navigating the undergraduate education system in far more complex ways than previously. Increasingly, students, including those seeking STEM degrees, are combining credits from multiple institutions to earn a degree, are transferring from 2-year to 4-year institutions (often without completing a degree or certificate program), are

159

transferring from 4-year to 2-year institutions, are enrolling at multiple institutions both simultaneously and sequentially, and are taking college credit in high school through dual enrollment and advanced placement courses (see Eagan et al., 2014; Salzman and Van Noy, 2014; Van Noy and Zeidenberg, 2014).

In the face of these changes in the student population, the committee found that—although there are some notable exceptions—postsecondary institutions, STEM departments, accrediting entities, and state and federal education policy have been slow to adapt. Although there are many small- and larger-scale efforts to remove the barriers that students face, we find that the underlying causes of these barriers need to be addressed much more deeply and systematically for widespread and sustainable reform to take hold. An important reason that institutions of higher education struggle to consistently deliver high-quality education experiences for STEM aspirants is that the institutions themselves and undergraduate education more generally were designed to serve much different student populations and to help them progress along much different education pathways than are typically being used today. In a sense, higher education institutions function more like a collection of discrete practices and policies, rather than being interconnected and synergistic.

There are many examples of unchanged policies and programs:

- a "weed-out" culture in many STEM departments rather than a supportive environment;
- graduation rates that are tracked on a 2-, 4-, or 6-year time clock, uninformed by data on median time to degree for different fields or the need to account for remediation time or the reality of part-time study;
- recognition and rewards to institutions for the quantity of degrees awarded rather than the quality, relevance, and levels of learning that are expected of and provided to students; and
- completion rates that are calculated on the basis of enrollment by first-time, full-time students and so discount part-time students and transfer students.

Several facts are worth noting. Institutions that take on the challenge of providing a high-quality STEM education to students from disadvantaged backgrounds often do so with fewer resources than elite institutions. Underrepresented minority students and first-generation students are more likely to enroll at a 2-year institution than a 4-year institution (Van Noy and Zeidenberg, 2014). Historically black colleges and universities award about 20 percent of all of the STEM bachelor's degrees earned by black students in fields other than psychology and social sciences, and about one-third of black students who have earned a Ph.D. in these STEM fields attained a bachelor's degree in STEM from historically black colleges and universities (National Science Foundation, 2013).

Two overarching findings undergird our conclusions and recommendations:

1. The "STEM pipeline" metaphor focuses on the students who enter at one end of the education system and those who emerge with STEM degrees. The metaphor does not reflect the diverse ways that students now move across and within higher education institutions, the diversity of paths that lead students to STEM degrees, or the expanding range of careers for those with STEM degrees. The "STEM pathways" metaphor is a more comprehensive and inclusive way of examining how students progress through STEM degrees and the much broader kinds of supports that higher education needs to provide to enable these students to successfully complete a credential.

2. Undergraduate STEM reform efforts have been piecemeal and not institutional in nature, and those that do not attend to today's students, their challenges or to the policy environments in which the institutions operate are likely to be short-lived and largely ineffective.

In the following three sections, we present our conclusions and recommendations related to today's students, about the role of institutions in serving those students, and about the need for systemic and sustainable change. Our conclusions and recommendations are embedded in these sections. In addition, our recommendations are presented by stakeholder group in Box 7-1.

TODAY'S STEM STUDENTS

CONCLUSION 1 There is an opportunity to expand and diversify the nation's science, technology, engineering, and mathematics (STEM) workforce and STEM-skilled workers in all fields if there is a commitment to appropriately support students through degree completion and provide more opportunities to engage in high-quality STEM learning and experiences.

Interest in STEM degrees among all undergraduate degree seekers at 2-year and 4-year institutions is at an all-time high, including students from traditionally underrepresented groups. Interest in STEM degrees is not only reflected in what degrees students indicate they are most interested in

BOX 7-1
Recommendations by Actors

STEM Departments and Academic Units

RECOMMENDATION 9 Disciplinary departments, institutions, university associations, disciplinary societies, federal agencies, and accrediting bodies should work together to support systemic and long-lasting changes to undergraduate science, technology, engineering, and mathematics education.

Colleges and Universities

RECOMMENDATION 1 Data collection systems should be adjusted to collect information to help departments and institutions better understand the nature of the student populations they serve and the pathways these students take to complete science, technology, engineering, and mathematics degrees.

RECOMMENDATION 4 Institutions, states, and federal policy makers should better align educational policies with the range of education goals of students enrolled in 2-year and 4-year institutions. Policies should account for the fact that many students take more than 6 years to graduate, and should reward 2-year and 4-year institutions for their contributions to the educational success of students they serve, which includes not only those who graduate.

RECOMMENDATION 5 Institutions of higher education, disciplinary societies, foundations, and federal agencies that fund undergraduate education should focus their efforts in a coordinated manner on critical issues to support science, technology, engineering, and mathematics (STEM) strategies, programs, and policies that can improve STEM instruction.

RECOMMENDATION 6 Accrediting agencies, states, and institutions should take steps to increase the alignment of policies that can improve the transfer process for students.

RECOMMENDATION 8 Institutions should consider how expanded and improved co-curricular supports for science, technology, engineering, and mathematics (STEM) students can be informed by and integrated into work on more systemic reforms in undergraduate STEM education to more equitably serve their student populations.

RECOMMENDATION 9 Disciplinary departments, institutions, university associations, disciplinary societies, federal agencies, and accrediting bodies should work together to support systemic and long-lasting changes to undergraduate science, technology, engineering, and mathematics education.

States and Federal Agencies

RECOMMENDATION 1 Data collection systems should be adjusted to collect information to help departments and institutions better understand the nature of the student populations they serve and the pathways these students take to complete science, technology, engineering, and mathematics degrees.

RECOMMENDATION 2 Federal agencies, foundations, and other entities that fund research in undergraduate science, technology, engineering, and mathematics (STEM) education should prioritize research to assess whether enrollment mobility in STEM is a response to financial, institutional, individual, or other factors, both individually and collectively, and to improve understanding of how student progress in STEM in comparison with other disciplines is affected by enrollment mobility.

RECOMMENDATION 3 Federal agencies, foundations, and other entities that support research in undergraduate science, technology, engineering, and mathematics education should support studies with multiple methodologies and approaches to better understand the effectiveness of various co-curricular programs.

RECOMMENDATION 4 Institutions, states, and federal policy makers should better align educational policies with the range of education goals of students enrolled in 2-year and 4-year institutions. Policies should account for the fact that many students take more than 6 years to graduate, and should reward 2-year and 4-year institutions for their contributions to the educational success of students they serve, which includes not only those who graduate.

RECOMMENDATION 5 Institutions of higher education, disciplinary societies, foundations, and federal agencies that fund undergraduate education should focus their efforts in a coordinated manner on critical issues to support science, technology, engineering, and mathematics (STEM) strategies, programs and policies that can improve STEM instruction.

RECOMMENDATION 6 Accrediting agencies, states, and institutions should take steps to increase the alignment of policies that can improve the transfer process for students.

RECOMMENDATION 7 State and federal agencies and accrediting bodies together should explore the efficacy and tradeoffs of different articulation agreements and transfer policies.

RECOMMENDATION 9 Disciplinary departments, institutions, university associations, disciplinary societies, federal agencies, and accrediting bodies should work together to support systemic and long-lasting changes to undergraduate science, technology, engineering, and mathematics education.

continued

BOX 7-1 Continued

Foundations

RECOMMENDATION 2 Federal agencies, foundations, and other entities that fund research in undergraduate science, technology, engineering, and mathematics (STEM) education should prioritize research to assess whether enrollment mobility in STEM is a response to financial, institutional, individual, or other factors, both individually and collectively, and to improve understanding of how student progress in STEM in comparison with other disciplines is affected by enrollment mobility.

RECOMMENDATION 3 Federal agencies, foundations, and other entities that support research in undergraduate science, technology, engineering, and mathematics education should support studies with multiple methodologies and approaches to better understand the effectiveness of various co-curricular programs.

RECOMMENDATION 4 Institutions, states, and federal policy makers should better align educational policies with the range of education goals of students enrolled in 2-year and 4-year institutions. Policies should account for the fact that many students take more than 6 years to graduate, and reward 2-year and 4-year institutions for their contributions to the educational success of students they serve, which includes not only those who graduate.

RECOMMENDATION 5 Institutions of higher education, disciplinary societies, foundations, and federal agencies that fund undergraduate education should focus their efforts in a coordinated manner on critical issues to support science, technology, engineering, and mathematics (STEM) strategies, programs, and policies that can improve STEM instruction.

---

earning when they first begin their undergraduate studies, but also in the fact that one-third of students who begin with an undeclared major select a STEM discipline as a major (Eagan et al., 2014).

The degree completion rates for all STEM aspirants is less than 50 percent, with the lowest completion rates found among students from underrepresented groups (blacks, Hispanics, and Native Americans). Three common threads among students from groups with low degree completion rates are that they have the greatest economic need, are more likely to require developmental courses, and have few if any immediate family members who completed college. Increasingly, students who aspire to earn STEM degrees are coming to college with a broad range of life experiences, are transferring among institutions at least once, and are more frequently stopping out. They are also likely to be working while attending college, especially 2-year colleges, and some are parents. Although the demographic

Disciplinary Professional Membership Organizations

RECOMMENDATION 5 Institutions of higher education, disciplinary societies, foundations, and federal agencies that fund undergraduate education should focus their efforts in a coordinated manner on critical issues to support science, technology, engineering, and mathematics (STEM) strategies, programs, and policies that can improve STEM instruction.

Accrediting Bodies

RECOMMENDATION 6 Accrediting agencies, states, and institutions should take steps to increase the alignment of policies that can improve the transfer process for students.

RECOMMENDATION 9 Disciplinary departments, institutions, university associations, disciplinary societies, federal agencies, and accrediting bodies should work together to support systemic and long-lasting changes to undergraduate science, technology, engineering, and mathematics education.

University Associations and Organizations

RECOMMENDATION 9 Disciplinary departments, institutions, university associations, disciplinary societies, federal agencies, and accrediting bodies should work together to support systemic and long-lasting changes to undergraduate science, technology, engineering, and mathematics education.

composition of students who are seeking STEM degrees is shifting, it remains true that on average, STEM aspirants arrive on campus better prepared and having achieved more academically than the student body as a whole. Yet only 40 percent of these students earn STEM degrees within 6 years.

Students who enter college declaring that they are interested in pursuing STEM degrees but then decide to enroll in non-STEM majors most frequently do so after STEM introductory courses (or prerequisite introductory science and mathematics courses). These students turn away from STEM in response to the teaching methods and atmosphere they encountered in STEM classes (President's Council of Advisors on Science and Technology, 2012; Seymour and Hewitt, 1997). Furthermore, many students who switch majors after their experiences in introductory STEM courses pass those courses. It seems that they abandon their goal of earning a STEM degree due to the way that STEM is taught and the difficulty in attaining support. That support, such as tutoring, mentoring, authentic STEM experiences, or other supports, improves retention in STEM majors (Estrada, 2014). In other words, students are dissuaded from studying STEM rather than being drawn into studying a different discipline. While some of the switching may be the result of considered choices based on opportunities to explore attractive alternatives, lack of a supportive environment in STEM likely contributes to those decisions.

Based on STEM persistence and completion rates, and research on why students leave, it seems clear that 2-year and 4-year institutions are not consistently providing all STEM degree seekers with a high-quality education experience and the supports that they need to succeed, especially in introductory and gateway courses.

CONCLUSION 2 Science, technology, engineering, and mathematics (STEM) aspirants increasingly navigate the undergraduate education system in new and complex ways. It takes students longer for completion of degrees, there are many patterns of student mobility within and across institutions, and the accommodation and management of student enrollment patterns can affect how quickly and even whether a student earns a STEM degree.

An increasing percentage of STEM aspirants and those who graduate with a STEM degree or certificate begin their college career at 2-year institutions. This is especially true among black, Hispanic, and American Indian students. In addition, the rate at which STEM aspirants and graduates transfer from a 4-year institution to a 2-year institution (reverse transfer) is also increasing (Salzman and Van Noy, 2014). Likewise, there is increased availability of and enrollment in high school dual-enrollment programs and Advanced Placement and International Baccalaureate STEM courses, both of which provide students with college-level courses and are accepted for college credit and placement at many institutions. The increased movement of undergraduate STEM credential aspirants often leads to loss of credits earned (because some credits do not transfer), classes that may not count toward the degree requirements in a second institution, and difficulties in adjusting to new academic cultures. All of these factors influence the amount of time it takes STEM aspirants to graduate, even if they are consistently making progress toward their degree and doing well in their classes. Students who reverse transfer (from a 4-year to a 2-year institution) are substantially less likely to complete a STEM degree within 6 years. However, students who concurrently enroll in multiple institutions are only slightly less likely to complete a STEM degree in 6 years than those who attend only one institution. Students who need remedial classes also need to take more credits, which often extends their time to graduation and increases the cost of their education. This is one reason that students with remedial needs often "time out" of federal financial aid.

CONCLUSION 3 National, state, and institutional undergraduate data systems often are not structured to gather information needed to understand how well the undergraduate education system and institutions of higher education are serving students.

Most large-scale data systems that include information on undergraduate students were built to track students in a pipeline model. Some systems focus primarily on gathering data on full-time or first-time students, while others do not account well for the swirling of students among institutions. These systems often rely on graduation rates as the sole metric of success for students and institutions: they do not systematically collect information on students' goals, reasons for transferring or leaving institutions, progress toward a credential, nor do they provide access to evidence-based teaching practices or student support systems.

The limitations of the data systems make it difficult for the states and the federal government to understand how the postsecondary education system is serving students, if some students are being served better than others, and which institutions consistently do not meet the needs of their students. In addition, most faculty, departments, and institutions do not know when students encounter barriers to earning the degree they seek or what supports students may need to succeed.

RECOMMENDATION 1 Data collection systems should be adjusted to collect information to help departments and institutions better understand the nature of the student populations they serve and the pathways these students take to complete science, technology, engineering, and mathematics (STEM) degrees.

- Colleges and universities need to more consistently leverage the information collected across their campuses (e.g., offices of institutional research, STEM departments, and student aid offices) to better understand who their students are, their movement among majors and institutions, the barriers they encounter in working toward their degrees, and the services or supports they need.
- States and federal agencies should consider how the information they require institutions to collect might enable better tracking of students through pathways they take to earn a STEM degree within and especially across institutions. In addition, they should consider

expanding measures of success, which increasingly inform funding formulas, beyond graduation rates.

There are a growing number of institutions that are using the data collected across their institutions to support student learning and identify when and where students need support to continue with their work toward STEM degrees. More campuses are identifying difficult introductory courses to provide supplemental instruction or use evidence-based instructional strategies and track students with data dashboards to improve progress toward degrees; however, systematic collection and use of such data are not widespread. With a better understanding of what barriers students typically encounter, and when and why students typically encounter them, institutions can more efficiently provide individualized support to students.

Existing data on undergraduate students and institutions are limited in a number of ways. We were not able to ascertain the success of STEM students who transferred from community colleges without earning a credential, nor could we address questions related to what happens to students who "time out" of financial aid.

A vision of success that goes beyond graduation rates and time to completion has been emerging from definitions of success developed by various stakeholder groups, including the American Association of Community Colleges, the Aspen Institute, the Bill & Melinda Gates Foundation, the National Governors Association, and the Association of American Universities. These stakeholders have identified a broad set of academic indicators, such as success in remedial and first-year courses, course completion, credit accumulation, credits to degree, retention and transfer rates, degrees awarded, expanding access, and learning outcomes. Much work is needed by these and other stakeholders to develop a systematic, national data source on such factors.

> RECOMMENDATION 2 Federal agencies, foundations, and other entities that fund research in undergraduate science, technology, engineering, and mathematics (STEM) education should prioritize research to assess whether enrollment mobility in STEM is a response to financial, institutional, individual, or other factors, both individually and collectively, and to improve understanding of how student progress in STEM, in comparison with other disciplines, is affected by enrollment mobility.

Many students move across institutions and into and out of STEM programs; the incidence is higher among community college students. It is often not clear what drives their decisions. One-half of community college STEM students enter into STEM after their first year of enrollment, and little is known about what factors are involved in their decisions and the ultimate implications for student outcomes. While late decisions can force students to take more than the required number of credits for a major because many STEM programs are highly structured with various requirements, early decisions may not be possible or even desirable if students are unsure about their career paths and need time to discover their interests. These decisions may be influenced by institutional policies (e.g., on early deadlines to declare program entry), discipline-based professional societies, and accrediting bodies. Research is needed on:

- what kinds of exploration students undertake as they decide to major (or not) in a STEM field and how they make their decisions,
- why students enter STEM programs at different times,
- the factors that attract them to STEM majors,
- how institutional structures might facilitate or delay their entry into STEM, and
- to what extent the identified problems may be associated with changing student demographics.

INSTITUTIONAL SUPPORT FOR TODAY'S STEM STUDENTS

CONCLUSION 4 Better alignment of science, technology, engineering, and mathematics (STEM) programs, instructional practices, and student supports is needed in institutions to meet the needs of the populations they serve. Programming and policies that address the climate of STEM departments and classrooms, the availability of instructional supports and authentic STEM experiences, and the implementation of effective teaching practices together can help students overcome key barriers to earning a STEM degree, including time to degree and the price of a STEM degree.

Substantial research in the last decade indicates that persistence in STEM is related to a host of factors that go beyond academic preparation of the individual student. Those factors include institutional practices and supports that reinforce student identities as scientists or engineers, recognition of talent, interaction with peers, and opportunities for authentic research experiences. Instructional practices that encourage active and interactive learning are keys to improving student learning and persistence in STEM. In addition, faculty behavior and attitudes inside and outside the classroom can provide cues that help students persist toward STEM degrees.

Discipline-Based Education Research (National Research Council, 2012) identifies the evidence-based practices that improve student learning and persistence in STEM programs. The study illustrates the importance of active instructional practices that engage students in the learning process and increase their interaction with peers, faculty, and teaching assistants. The report also points to the slow adoption of these practices. Research has also shown increased effects of evidence-based teaching practices when paired with co-curricular supports.

Even when high-quality instructional practices are implemented, students often receive little guidance or support regarding how efficiently to navigate the vast array of undergraduate education options. This makes it difficult for students to know how to get from where they are academically to where they want to be or to help them explore options that they have not considered about current and future career opportunities. This situation may help explain the phenomena of students who take classes at multiple institutions, transfer between institutions, or take time off from college, but all of this "churning" is associated with lower rates of completion and longer times to degree. Time is the enemy of many undergraduate STEM students. As time to degree increases, the likelihood of graduating seems to decrease due to a host of factors, perhaps, most importantly, increasing student debt.

Long-term program evaluations of interventions now provide evidence of what can increase persistence and graduation rates among STEM students. The most promising interventions combine contact with faculty and a supportive peer group along with access to authentic STEM experiences. Undergraduate research experiences show positive effects for both persistence and intentions for graduate school, over and above faculty mentoring experiences (though the two are often combined in structured research programs). Co-curricular supports (e.g., research experiences, mentoring, summer bridge programs, and living and learning communities) have been shown to affect STEM student persistence and completion when they align with evidence-based practices in supporting student learning and interests.

The culture of STEM classrooms and departments also influences STEM student persistence. Many students interested in STEM degrees, especially those from underrepresented groups and women, decide to pursue other fields due to the instructional practices, the "weed out" culture of some introductory STEM courses, and the lack of opportunities to engage in authentic STEM experiences.

To train effective mentors and create a culture of inclusiveness, faculty need to be provided opportunities to become more aware of implicit bias and stereotyping as well as how to avoid them. Departments need to encourage greater student involvement in research and design experiences, as well as in clubs and organizations related to a discipline, which have been shown to improve retention in STEM (Chang et al., 2014; Espinosa, 2011). The role of professional STEM clubs and organizations also points to the importance of local chapters as well as national student organizations and the development or enhancement of professional society programs for undergraduates to sustaining interest and retention in STEM.

The need for and nature of student supports likely will differ by type of institution and student background. It would be useful for institutional leaders to collect the kind of data about students' current interests and needs to better determine how they can offer a range of interventions that are most appropriate to the current and changing needs of their students.

In general, 2-year and 4-year institutions serve students with different backgrounds, goals, and educational preparation. Community colleges enroll more older, first-generation, and working students than 4-year colleges. They play a significant role in the pathways that a diverse population of students takes in earning STEM degrees and certificates. Science and engineering programs at 2-year institutions enrolled relatively high proportions of Hispanic, Asian, and female students but a lower proportion of black students, who were more likely to be enrolled in technical-level programs.

Although community college STEM students have relatively low completion rates, their high persistence rates are notable. Students who begin their undergraduate education at a 2-year institution often take more than 6 years to complete their degrees, due to part-time enrollment, interruptions in their enrollment, and loss of course credit when they transfer between institutions. Understanding the quality of the educational experiences provided by 2-year institutions is hampered by the existing data systems that do not provide clear information on students who transfer from 2-year institutions to 4-year institutions without earning a degree or certificate. In addition, the contribution of 2-year institutions to the degrees that transfer students receive at 4-year institutions is not tracked and so is not well understood. Although there is emerging evidence regarding the characteristics of departments that support the use of evidence-based pedagogy, we were unable to find data on the relative use of such pedagogy. In fact, we were unable to even find recent national data on who teaches STEM courses (full-time tenured faculty, adjunct, or other), the level of instructional training that instructors had received, or alignment of instructor practices with evidence-based practices.

RECOMMENDATION 3 Federal agencies, foundations, and other entities that support research in undergraduate science, technology, engineering, and mathematics education should support studies with multiple methodologies and approaches to better understand the effectiveness of various co-curricular programs.

Future research on co-curricular programs should reflect the complexity and "messiness" of undergraduate education, and it should illuminate how the co-curricular support fits into the broader culture of institutions.

There is a need for more studies that track students over time to assess both the short-term and long-term effects of program elements across academic pathways. Such studies should include data from similar cohorts of students who do not participate in the program as a comparison or control group. When possible and appropriate, participants should be randomly assigned to co-curricular program groups.

For these studies to be useful, co-curricular programs need to identify measurable outcomes such as retention, grades, knowledge, and degree conferment, and they should identify the discipline of study. In-depth case studies or focus groups with program participants and similar students to track experiences at time of participation and shortly after can add to the research. Studies should move beyond linear models of student progress to a credential to test models that are more reflective of the kind of decision making of students. In addition, studies of long-time co-curricular programs and the nature of the sites that house them are needed to better understand how to sustain successful programs.

> **RECOMMENDATION 4 Institutions, states, and federal policy makers should better align educational policies with the range of education goals of students enrolled in 2-year and 4-year institutions. Policies should account for the fact that many students take more than 6 years to graduate, and should reward 2-year and 4-year institutions for their contributions to the educational success of students they serve, which includes not only those who graduate.**

- The states and the federal government should revise undergraduate accountability policies so that systems of assessment, evaluation, and accountability give credit to and do not penalize (i.e., in-state funding formulas) institutions for supporting students' progress toward their desired educational outcome. It is important that policies take into account the various ways that students are currently using different institutions in pursuit of a degree, certification, or technical skills.
- The states and the federal government should extend financial aid eligibility to graduation for students making satisfactory progress toward a degree or certificate, so that students do not "time out" of financial aid eligibility.
- Colleges and universities should shift their institutional policies toward a model in which all students who are admitted to a degree program are expected to complete that program and are provided the instruction, resources, and support they need to do so, rather than a model in which it is assumed that a large fraction of students will be unsuccessful and will leave science, technology, engineering,

and mathematics programs. This model can save money because the time to degree is shortened and the number of drops, failures, withdrawals, and repeating of courses is reduced.

Systems of accountability for undergraduate education need to better align to the pathways that students actually are taking to earn STEM degrees. To do so, more thought needs to go into how each institution can track students' progression toward a degree or other outcome—including gaining skills to upgrade current employment and earning a certificate while working toward an associate's degree—recognizing the long time to degree completion among many STEM students.

STEM students are taking longer to earn degrees because of many factors, including transferring among institutions, changing majors, and the need to follow strict course sequencing. It is now uncommon for a student to complete a 2-year degree in 2 years or a 4-year degree in 4 years. The time frame of some current financial aid policies do not reflect what is now common and do not align with the pathways that students are taking to earn degrees. Providing financial aid on the basis of the number of semesters a student has spent in college has a differentially negative impact on students from underrepresented minority groups, who more frequently than other students need remedial courses due to weakness in their K-12 preparation, starting at 2-year institutions, and taking longer to graduate. Financial aid policies could recognize the current pathways by focusing on whether students are making adequate progress toward their academic goals.

The culture of many STEM courses and departments is undergirded by the belief that "natural" ability, gender, or ethnicity is a significant determinant of a student's success in STEM. Related to this view is the tendency for introductory mathematics and science courses to be used as "gatekeeper" or "weeder" courses, which may discourage students from pursuing STEM degrees, through highly competitive classrooms and a lack of pedagogy that promotes active participation and emphasizes mastery and improvement. These courses often seek to select out and distinguish those with some perceived ability in STEM. The classroom and departmental culture needs to value diversity and be based on the understanding that all students aspiring to earn a STEM degree have the potential to succeed in STEM and provide all students the opportunity to make an informed decision about whether they want to continue pursuing STEM credentials.

RECOMMENDATION 5 Institutions of higher education, disciplinary societies, foundations, and federal agencies that fund undergraduate education should focus their efforts in a coordinated manner on critical issues to support science, technology, engineering, and mathematics

(STEM) strategies, programs, and policies that can improve STEM instruction.

- Colleges and universities should adjust faculty reward systems to better promote high-quality instruction and provide support for faculty to integrate effective teaching strategies into their practice. They should encourage educators to learn about and implement effective teaching methods by supporting participation in workshops, professional meetings, campus-based faculty development programs, and other related opportunities. Instructional quality is a key aspect of a student's undergraduate experience that could be addressed by providing incentives for more faculty members to align their classroom practices with evidence-based pedagogy.
- Disciplinary and professional membership organizations should become more active in developing tools to support evidence-based teaching practices, and providing professional development in using these active pedagogies for new and potential faculty members and instructors.
- The National Center for Education Statistics of the U.S. Department of Education should collect systematic data on tenured, tenure-track, and nontenure-track faculty and staff, as it previously did through the National Study of Postsecondary Faculty. Such data will make it possible to understand who is teaching STEM courses and whether they have participated in professional development programs to implement evidence-based instructional strategies. The Department of Education should support research on what supports are needed to allow all educators, including tenured, tenure-track faculty, full-time nontenured teaching faculty, adjunct faculty, and lecturers, to successfully implement such strategies.
- Federal agencies, foundations, and other entities should invest in implementation research to better understand how to increase adoption of evidence-based instructional strategies.

Although a considerable body of research is emerging about the nature and effect of effective instructional practices, this awareness has not necessarily been translated into widespread implementation of such practices in STEM classrooms. More investment needs to be made in implementation research to determine how to support putting this knowledge into practice. There have been calls for working with postdoctoral scholars and graduate students during their education to ensure that professional development is available to them on effective teaching strategies. This requires departmental support and leadership across an institution, along with agreement that future faculty should have mastered research-based teaching strategies as well as disciplinary research knowledge and skills.

RECOMMENDATION 6 Accrediting agencies, states, and institutions should take steps to support increased alignment of policies that can improve the transfer process for students.

- Regional accrediting bodies should review student outcomes by participating colleges and require periodic updates of articulation agreements in response to those student outcomes.
- States should encourage tracking transfer credits and using other metrics to measure the success of students who transfer.
- Colleges and universities should work with other institutions in their regions to develop articulation agreements and student services that contribute to structured and supportive pathways for students seeking to transfer credits.

The pathways that students are taking to earn undergraduate STEM degrees have become increasingly complex, with greater numbers of students earning credits at more than one institution. Thus, issues of transfer and articulation are now relevant to an increasing proportion of STEM students, as well as students in other majors. The range of different regional, state, and institutional transfer and articulation policies that students encounter can be dizzying, and they can extend a student's time to completion and increase the cost of college, as well as being stressful to navigate.

Regional accrediting agencies, states, and institutions can all take steps to remove the barriers that students currently face when transferring credits among institutions. Removing these barriers may require creative and collaborative solutions, but they have the potential not only to improve students' educational experience, but also to make higher education institutions more efficient and effective.

RECOMMENDATION 7 State and federal agencies and accrediting bodies together should explore the efficacy and tradeoffs of different articulation agreements and transfer policies.

There is a need to better understand the efficacy of existing and new models of articulation agreements. Currently, it is not clear which types of agreements work for different types of students (including students from underrepresented groups and veterans), and for different types of transfers (vertical, reverse, and lateral). Research on the effects of articulation agreements needs to consider not only the policies that guide the transfer of credits, but also the supports developed to make it easier for students to navigate the policies and adjust to their different academic environments.

SYSTEMIC AND SUSTAINABLE CHANGE IN STEM EDUCATION

CONCLUSION 5 There is no single approach that will improve the educational outcomes of all science, technology, engineering, and mathematics (STEM) aspirants. The nature of U.S. undergraduate STEM education will require a series of interconnected and evidence-based approaches to create systemic organizational change for student success.

From years of attempts to improve higher education for all, many lessons have been learned. Focusing narrowly on individuals rather than on the entire system is not effective because it leads to changes of minimal scale and sustainability. Failing to leverage the many actors in education—individuals, departments, institutions, disciplinary societies, business and industry, governments—in a systematic fashion is ineffective because different levels of the education system often operate in isolation and are often unaware of how their actions can both affect and be affected by other components of the system.

In addition, focusing narrowly on pedagogical and curricular changes and not considering other variables that are related to student success, such as institutional policies, articulation, faculty culture, and financial aid, limits the potential effects of such changes. It is not productive to focus on "silver bullets": they often lead to "fixing the student" approaches rather than identifying problems throughout the system, from mathematics preparation, to science culture, to faculty teaching, to financial aid, to articulation and transfer. Finally, it is clear that such barriers to change as the nature of the incentive structure in colleges and universities remain largely unaddressed, and studies have not been conducted to determine if addressing such barriers would facilitate large-scale and sustainable change in institutions or education systems.

CONCLUSION 6 Improving undergraduate science, technology, engineering, and mathematics education for all students will require a more systemic approach to change that includes use of evidence to support institutional decisions, learning communities and faculty development networks, and partnerships across the education system.

Students need a higher education system that is less fragmented—or at least has clearer road markers—so that the diverse and complex pathways they take toward a degree do not create unnecessary barriers. Partnerships with elementary and secondary schools may be able to lead to better preparation for college, especially in mathematics. Partnerships with employers can lead to better articulation of the skills and knowledge that are relevant for their workforces, as well as opportunities for internships and work-related experiences that may improve students' understanding of and commitment to STEM education.

At the institutional level, program faculty and administrators need to recognize that successful improvements usually include strong leadership, including support for faculty to undertake the changes needed; awareness of how to overcome the barriers to adaptation and implementation of curricula that have been demonstrated to be effective; faculty who implement instructional practices developed through discipline-based education research; and data to monitor students' progress and to hold departments accountable for losses and recognize and reward them for student success.

Strong, multi-institutional articulation agreements, including common general education, common introductory courses, common course numbering, and online, easily available student access to equivalencies, can improve the percentage of contributory credits transferred, shorten the time to degree, and increase completion rates.

Department-level leadership is critical for systematic change. It can drive changes in rigid course sequencing requirements, transfer credit policies, degree requirements, differential tuition policies, and classroom practices. It can build connections between the reform efforts in their department and broader efforts in their institutions, as well as connect to multi-institutional reform efforts supported by foundations and disciplinary associations. The training of STEM department chairs supported by a number of programs and professional organizations has yielded promising results for departmental programs and their students.

> **RECOMMENDATION 8 Institutions should consider how expanded and improved co-curricular supports for science, technology, engineering, and mathematics (STEM) students can be informed by and integrated into work on more systemic reforms in undergraduate STEM education to more equitably serve their student populations.**

To improve degree attainment rates, the quality of programs, and better serve their diverse student populations, institutions can consider a wide range of policies and programs: initiating or increasing opportunities for undergraduate student participation in research and other authentic STEM experiences; connecting students to experiences related to careers in their field of interest; expanding the use of educational technologies that have been effective in addressing the remediation needs of students; building student learning communities; and providing access to college and career guidance to help students understand the various and most efficient pathways to the degrees and careers they want. Students seem to benefit most from such supports when they are paired with evidence-based instructional strategies and when three or more co-curricular supports are bundled together (Estrada, 2014). Such efforts will be more sustainable and effective if they are integrated into more systemic reform efforts.

> **RECOMMENDATION 9 Disciplinary departments, institutions, university associations, disciplinary societies, federal agencies, and accrediting bodies should work together to support systemic and long-lasting changes to undergraduate science, technology, engineering, and mathematics education.**

- STEM departments and entire academic units should support learning communities and networks that can help change faculty belief systems and practices and develop sustainable changes.
- Colleges and universities should offer instructor training and mentoring to graduate students and postdoctoral scholars. Participating in such efforts as The Center for the Integration of Research, Teaching, and Learning (funded by the National Science Foundation; see Chapter 3) can educate graduate students about the value of treating their teaching as a form of scholarship and to value use of evidence-based approaches to teaching.
- University associations and organizations should continue to facilitate undergraduate STEM educational reforms in their member institutions, particularly by examining reward structures and barriers to change and providing resources for data collection on student success, as well as by providing resources for interventions, support programs, and ways to share effective practices.
- Disciplinary societies should support the development of continuing and intensive national and regional faculty development programs, awards, and recognition to encourage use of evidence-based instructional practices.
- Federal agencies that support undergraduate STEM education should consider giving greater priority to supporting large-scale transformation strategies that are conceptualized to include and extend beyond instructional reform, and they should support both implementation research and research on barriers to reform that can support success for all students. They should increase the percentage of undergraduate STEM reform efforts and projects that focus on multiple levels—department, institution, discipline, government, and business and industry.
- Following the policies adopted by some disciplinary accrediting bodies (e.g., the Accreditation Board for Engineering and Technol-

ogy), regional and professional accrediting bodies should consider incorporating evidence-based instructional practices and faculty professional development efforts into their criteria and guidelines.

The nature of the challenges of removing the barriers to 2-year and 4-year STEM degree completion can only be addressed by a system of solutions that includes the commitment to transformation. Looking from the ground up, those who teach need to be enabled to adopt and engage in effective classroom practices; co-curricular supports need to be made available for students who begin college with interest in STEM but who may lack some of the skills necessary to be immediately successful in their pursuit of study in STEM.

Money still matters: strategies need to be explored for addressing financial need in ways that connect students to STEM (such as through STEM-related work-study programs and internships and co-ops) rather than distracting them from it. Providing quality advice about courses, fields of study, careers, and navigating the many college pathways in STEM—as well as supporting learning communities—can help avoid many of the pitfalls that can delay or prevent degree completion.

Looking across institutions, the policy barriers to articulation and alignment need to be addressed. Although some removal of barriers can be promoted locally through, for example, the active commitment of individuals, (e.g., chemistry faculty in 4-year institutions working directly with chemistry faculty in feeder 2-year institutions and high schools), a negatively structured policy environment can impede such interventions. There is a clear need to explore all the policy impediments that make navigation of the pathways to STEM degrees in and across institutional boundaries especially difficult, and there are examples in various states and institutions that can be considered to smooth STEM pathways.

Looking from the top down, leadership is needed at every level to support change. Institutional leaders need to be committed to providing the supportive infrastructure that can make grassroots pedagogical and administrative changes possible (including active classrooms, technology, co-curricular supports, data systems, and teaching-learning centers). Loss of state support has negatively affected the operational model of many public institutions, forcing increased costs to be passed through to students, which disproportionately affects those who can least afford to attend, extending time to degree and may affect students' choices of major (e.g., when there is differential tuition for programs such as engineering). National accountability structures, though well intentioned, currently reward the most selective institutions while penalizing those with fewer resources, but the latter are the ones who often enroll and succeed in enrolling STEM students from disadvantaged and less selective backgrounds. The admonishment to "first, do no harm" should lead to a national discussion of how to recognize and honor the work of such institutions. At the same time, highly resourced institutions can be challenged to better support their STEM students through programs of active retention rather than "weeding out."

Finally, leadership is required from all constituents, including state and federal government, funders, business and industry, and both higher education and STEM professionals, both within and across those communities. Rather than relying on failed or unsustainable structures that serve only a few or push out students who aspire to and are capable of completing a STEM degree, they should seek solutions that connect the pathways to STEM degrees.

REFERENCES

Chang, M., Sharkness, J., Hurtado, S., and Newman, C. (2014). What matters in college for retaining aspiring scientists and engineers from underrepresented racial groups. *Journal of Research in Science Teaching, 51*(5), 555–580.

Eagan, K., Hurtado, S., Figueroa, T., and Hughes, B. (2014). *Examining STEM Pathways among Students Who Begin College at Four-Year Institutions.* Commissioned paper prepared for the Committee on Barriers and Opportunities in Completing 2- and 4-Year STEM Degrees, National Academy of Sciences, Washington, DC. Available: http://sites.nationalacademies.org/cs/groups/dbassesite/documents/webpage/dbasse_088834.pdf [April 2015].

Espinosa, L.L. (2011). Pathways and pipelines: Women of color in undergraduate STEM majors and the college experiences that contribute to persistence. *Harvard Educational Review, 81*(2), 209–241.

Estrada, M. (2014). *Ingredients for Improving the Culture of STEM Degree Attainment with Co-curricular Supports for Underrepresented Minority Students.* Paper prepared for the Committee on Barriers and Opportunities in Completing 2- and 4-Year STEM Degrees. http://sites.nationalacademies.org/cs/groups/dbassesite/documents/webpage/dbasse_088832.pdf [April 2015].

National Center for Education Statistics (2013). *Digest of Education Statistics 2013.* Washington, DC: U.S. Department of Education.

National Research Council. (2012). *Discipline-Based Education Research: Understanding and Improving Learning in Undergraduate Science and Engineering.* Committee on the Status, Contributions, and Future Directions of Discipline-Based Education Research. S. Singer, N.R. Nielsen, and H.A. Schweingruber (Eds.). Board on Science Education, Division of Behavioral and Social Sciences and Education. Washington DC: The National Academies Press.

National Science Board. (2014). *Science and Engineering Indicators 2014.* NSB #14-01. Arlington VA: National Science Foundation.

National Science Foundation and National Center for Science and Engineering Statistics. (2013). *Women, Minorities, and Persons with Disabilities in Science and Engineering: 2013.* Arlington, VA: National Science Foundation.

President's Council of Advisors on Science and Technology. (2012). *Report to the President. Engage to Excel: Producing One Million Additional College Graduates with Degrees in Science, Technology, Engineering and Mathematics.* Available: http://www.whitehouse.gov/sites/default/files/microsites/ostp/pcast-engage-to-excel-final_feb.pdf [April 2015].

Salzman, H., and Van Noy, M. (2014). *Crossing the Boundaries: STEM Students in Four-Year and Community Colleges*. Paper prepared for the Committee on Barriers and Opportunities in Completing 2- and 4-Year STEM Degrees. Available: http://sites.nationalacademies.org/cs/groups/dbassesite/documents/webpage/dbasse_089924.pdf [April 2015].

Seymour, E., and Hewitt, N. (1997). *Talking about Leaving: Why Undergraduates Leave the Sciences*. Boulder, CO: Westview Press.

Van Noy, M., and Zeidenberg, M. (2014). *Hidden STEM Knowledge Producers: Community Colleges' Multiple Contributions to STEM Education and Workforce Development*. Paper prepared for the Committee on Barriers and Opportunities in Completing 2- and 4-Year STEM Degrees. Available: http://sites.nationalacademies.org/cs/groups/dbassesite/documents/webpage/dbasse_088831.pdf [April 2015].

Appendix A

Instructional Resources, Online Curriculum Repositories, and Situational Barriers to Change

Individual faculty members, group, or departments that are considering changes to science, technology, engineering and mathematics (STEM) education should carefully consider whether to try to develop new curriculum from scratch, a very time-consuming multi-year undertaking, or to take advantage of existing research-based curriculum. One legacy of the investment by funding agencies in research-based pedagogies is curricula, as well as curriculum and publications on the barriers and opportunities associated with implementing and sustaining them. In some STEM disciplines, funding agencies and national STEM organizations have organized online repositories for those curricula. ComPADRE in physics (Mason, 2007) and CourseSource in biology (Wright et al., 2013) are examples of resources for faculty interested in identifying new curriculum.

Some of the more prominent curriculum reform groups have developed resources for both new and experienced users. These may include electronic mailing lists, websites, or central resources for users, as well as topic-specific workshops or meetings for users, such as POGIL, BIOquest, the Academy of Inquiry Based Learning (IBL), and Sencer. Faculty interested in adopting one of these more prominent curricula can take advantage of these resources.

National discipline-based organizations are also an important support for faculty interested in implementing new curricula. Such organizations, including the American Association of Physics Teachers, the Mathematical Association of America, the American Society for Engineering Education (ASEE), and the American Geophysical Union, have sponsored meetings and workshops that allow STEM educators to become more familiar with research-based STEM curricula in a particular discipline. Manduca (2008) documents the importance of these meetings to discipline-wide STEM reform. At the broadest level, organizations such as the National Center for Academic Transformation offer support for revised cross-disciplinary STEM curricula, particularly at the introductory course level.

For validated STEM curricula, there are concept inventories, such as the force concept inventory (Hestenes et al., 1992), the chemistry inventory (Mulford and Robinson, 2002; Epstein, 2013), civil and environmental engineering (Sengupta et al., 2013), and the calculus concept inventory (Epstein, 2007). While these inventories have served as a source of critique of current STEM education, they can also serve as a resource for STEM reformers in communicating about successful curricula (Libarkin, 2008). Data from such inventories can be useful for improving curricular implementations and communicating the status and successes of STEM reform efforts to institutional and cross-institutional stakeholders.

Many successful and sustained curricular changes make significant changes to "situational barriers," identified by Henderson and Dancy (2011). These changes may include new classrooms specifically for STEM group work, such as SCALE-UP[1] or STUDIO classrooms,[2] or significantly revised temporal course structures, such as the CLASP[3] model, or completely reworked class structures, such as interdisciplinary programs at Pomona (Copp et al., 2012) or the paradigms model (Manogue and Krane, 2003) at Oregon State; or new or reconfigured buildings, such as makerspaces and student design buildings, which many engineering schools now house. Some of these programs have been ongoing for a long time.

These programs suggest that there is a correlation between sites where STEM reform has been adopted and persisted over time and positive situational barriers that make it difficult to return to traditional lecture and laboratory approaches. Some research (Lasry et al., 2014) on a SCALE-UP implementation site suggests that the existence of a reformed curriculum with a barrier to reversion (in this case, the modified classroom and schedule) may lead faculty who are required to teach in reformed classes to reconsider their own teaching methods.

More research is needed to determine whether such "positive situational barriers" (schedule changes, physical changes to classroom, significant revision to the curriculum, required faculty curricular meetings) support the sustainability of STEM reform curriculum by presenting barriers to return to traditional lecture/lab instruction modes. It is also possible

---

[1] For more information, see http://www.ncsu.edu/per/scaleup.html [July 2015].

[2] For more information, see http://serc.carleton.edu/introgeo/studio/why.html [July 2015].

[3] For more information, see http://www.aps.org/units/fed/newsletters/spring2011/webb.cfm [July 2015].

that curriculum or programs that by design require faculty to regularly discuss the curriculum may lead not just to sustained reform, but to even greater innovation.

REFERENCES

Copp, N.H., Black, K., and Gold, S. (2012). Accelerated integrated science sequence: An interdisciplinary introductory course for science majors. *Journal of Undergraduate Neuroscience Education, 11*(1), A76–A81.

Epstein, J. (2007). Development and validation of the calculus concept inventory. In *Proceedings of the Ninth International Conference on Mathematics Education in a Global Community*, September 7–12. Available: http://bit.ly/bqKSWJ [February 2016].

Epstein, J. (2013). The calculus concept inventory—Measurement of the effect of teaching methodology in mathematics. *Notices of the AMS, 60*(8), 1018–1026. Available: http://www.ams.org/notices/201308/rnoti-p1018.pdf [February 2016].

Henderson, C., and Dancy, M.H. (2011). *Increasing the Impact and Diffusion of STEM Education Innovations*. A White Paper commissioned for the Characterizing the Impact and Diffusion of Engineering Education Innovations Forum, February 7–8, New Orleans, LA.

Hestenes, D., Wells, M., and Swackhammer, G. (1992). Force concept inventory. *The Physics Teacher, 30*(March), 141–158. Available: http://modeling.asu.edu/R&E/FCI.PDF [February 2016].

Lasry, N., Charles, E., and Whittaker, C. (2014). When teacher-centered instructions are assigned to student-centered classrooms. *Physical Review Special Topics Physics Education Research, 10*(1), 010116-1–010116-9. Available: http://journals.aps.org/prper/pdf/10.1103/PhysRevSTPER.10.010116 [February 2016].

Libarkin, J. (2008). *Concept Inventories in Higher Education Science*. A manuscript prepared for the National Research Council Promising Practices in Undergraduate STEM Education Workshop 2, Washington, DC, October 13-14. Available: http://sites.nationalacademies.org/cs/groups/dbassesite/documents/webpage/dbasse_072624.pdf [February 2016].

Manduca, C.A. (2008). *Working with the Discipline: Developing a Supportive Environment for Education*. Presented at the National Research Council's Workshop Linking Evidence to Promising Practices in STEM Undergraduate Education, Washington, DC. Available: http://sites.nationalacademies.org/cs/groups/dbassesite/documents/webpage/dbasse_072635.pdf [April 2015].

Manogue, C.A., and Krane, K.S. (2003). Paradigms in physics: Restructuring the upper level. *Physics Today, 56*, 53–58.

Mason, B. (2007). *Introducing ComPADRE. Forum on Education. 2007. Communities for Physics and Astronomy Digital Resources in Education*. Available: http://www.compadre.org [April 2015].

Mulford, D.R., and Robinson, W.R., (2002). An inventory for alternate conceptions among first-semester general chemistry students. *Journal of Chemical Education, 79*(6), 739–743. Available: http://modeling.asu.edu/ModChem_web/Evaluation/CCI-old/p739.pdf [February 2016].

Sengupta, S., Cunningham, J.A., Ergas, S.J., Goel, R.K., Ozalp, D., and Reed, T. (2013, June), *Development of a Concept Inventory for Introductory Environmental Engineering Courses*. Paper presented at 2013 American Society for Engineering Education Annual Conference, Atlanta, GA. Available: https://peer.asee.org/19426 [February 2016].

Wright, R., Bruns, P.J., and Blum, J.E. (2013). CourseSource: A new open-access journal of scholarly teaching. *The FASEB Journal*, 27(Meeting Abstract Supplement), lb242. Available: http://www.fasebj.org/cgi/content/meeting_abstract/27/1_MeetingAbstracts/lb242 [February 2016].

Appendix B

Biographical Sketches of Committee Members and Staff

SHIRLEY MALCOM (*Chair*) is head of the Directorate for Education and Human Resources Programs of the American Association for the Advancement of Science (AAAS). She serves on several boards, including the Heinz Endowments, Public Agenda, and Digital Promise. She is an honorary trustee of the American Museum of Natural History, a regent of Morgan State University, and a trustee of the California Institute of Technology. She is a member of the National Academy of Sciences and a recipient of its Public Welfare Medal. She is a fellow of the AAAS and of the American Academy of Arts and Sciences. She served on the National Science Board and on the President's Committee of Advisors on Science and Technology. She has a B.S. in zoology from the University of Washington, an M.A. in zoology from the University of California, Los Angeles, and a Ph.D. in ecology from Pennsylvania State University.

CYNTHIA J. ATMAN is the founding director of the Center for Engineering Learning & Teaching (CELT), a professor in Human Centered Design & Engineering, and the inaugural holder of the Mitchell T. and Lella Blanche Bowie Endowed Chair at the University of Washington (UW). Dr. Atman is co-director of the Consortium to Promote Reflection in Engineering Education, funded by the Leona M. and Harry B. Helmsley Charitable Trust. She was director of the NSF-funded Center for the Advancement of Engineering Education, a national research center that was funded from 2003-2010. Dr. Atman joined the UW in 1998 after 7 years on the faculty at the University of Pittsburgh. Her research focuses on engineering design learning, considering context in engineering design, and the use of reflec-

187

tion to support learning. She is a fellow of the American Association for the Advancement of Science (AAAS) and the American Society of Engineering Education (ASEE). Dr. Atman holds a Ph.D. in Engineering and Public Policy from Carnegie Mellon University.

GEORGE BOGGS is a clinical professor of higher education for the Roueche Graduate Center at National American University and an adjunct professor of higher education at San Diego State University. He is superintendent and president emeritus of Palomar College and president and chief executive officer emeritus of the American Association of Community Colleges. Previously, he served as a faculty member, division chair, and associate dean of instruction at Butte College in California. He has served on the boards of the American Council on Education, the Educational Testing Service, the National Center for Postsecondary Research, the National Center for Community College Student Engagement, the National Science Foundation, and the Accrediting Board for Engineering and Technology. He has a B.S. in chemistry from Ohio State University, an M.S. in chemistry from the University of California at Santa Barbara, and a Ph.D. in education administration from the University of Texas, Austin.

PAMELA BROWN is associate provost at New York City College of Technology (City Tech) of the City University of New York, where she previously served as the dean of the School of Arts & Sciences. She also previously served as a program director in the Division of Undergraduate Education at the National Science Foundation (NSF). Her work has focused on creating initiatives to improve the retention and recruitment of students interested in careers in science, technology, engineering, and math fields. She was the principal investigator of an NSF project, "Metropolitan Mentors: (MMNet): Growing an Urban STEM Talent Pool across New York City." She is the first woman to have received a Ph.D. in chemical engineering from Polytechnic University (now NYU Polytechnic School of Engineering).

PETER BRUNS is a professor emeritus of genetics at Cornell University, where he held a number of positions, including professor of genetics, associate director of the Biotechnology Program, and director of the Division of Biological Sciences. Previously, he held several positions at the Howard Hughes Medical Institute. His scientific work includes pioneering methods to genetically manipulate the separate somatic and germinal nuclei of the single celled organism *Tetrahymena thermophila*. His work has also focused on education, including as founder of the Cornell Institute for Biology Teachers. He is the recipient of the Elizabeth W. Jones Award for Excellence in Education from the Genetics Society of America and of the Bruce Alberts Award for Excellence in Science Education from the American Society for

Cell Biology. He has an A.B. in zoology from Syracuse University and a Ph.D. in cell biology from the University of Illinois.

TABBYE CHAVOUS is a professor and associate dean for academic programs and initiatives at the Rackham Graduate School at the University of Michigan. Her research centers around racial and gender identity development among African American adolescents and young adults and their relationship with students' academic identities and its implications for academic and psychological adjustment. She also studies transitions to secondary schooling and higher education among ethnic minority students and racial and multicultural climates within secondary and higher education settings. She is a principal investigator and co-director of the university's Center for the Study of Black Youth in Context and is working on a project examining academic identification processes among African American students pursuing academic pathways in STEM fields. She has a Ph.D. in community psychology from the University of Virginia.

CHARLES DE LEONE is professor of physics at California State University at San Marcos, where he helped found the university's Committee on Undergraduate Research. He is principal investigator and director of a joint project with Palomar College aimed at increasing the number of graduates in science, technology, engineering, and mathematics fields. He has been the principal investigator and co-principal investigator on multiple studies aimed at developing, adapting, and implementing best practices science curriculum. His physics education research includes the areas of multiple representations, student use of technology, and the efficacy of active-learning-based pedagogy. He has worked as a consultant and leader in professional development programs nationwide. He has a B.S. in physics from Santa Clara University and a Ph.D. in physics from the University of California at Davis.

CATHERINE DIDION (*Senior Program Officer*) is on the staff of the National Academy of Engineering. Her portfolio includes projects on engineering education, the technical workforce, and diversity. Previously, she served as the executive director of the Association for Women in Science, and she has collaborated with the European Commission, the South African Ministry of Science and Technology, the Organization of American States, the InterAmerican Network of Academies of Sciences, and UNESCO. Her work focuses on issues of education, the workforce, and equity in engineering and science. In 2012 she was named one of 100 Women Leaders in STEM by STEM Connector. She completed her undergraduate degree at Mount Holyoke College and graduate work at the University of Virginia.

FRANK DOBBIN is a professor of sociology in the Department of Sociology and the Edmond J. Safra lab fellow at Harvard University. Previously, he was an assistant professor at Princeton University. He studies organizations, inequality, economic behavior, and public policy and explores institutional factors that affect the participation of minorities and women in science, technology, engineering, and mathematics. He has won the American Sociological Association's distinguished scholarly book award, the Max Weber award for inventing equal opportunity, and the Rosabeth Moss Kanter award for excellence in work-family research for civil rights law at work. He has a B.A. in sociology from Oberlin College and a Ph.D. in sociology from Stanford University.

MICHAEL FEDER (*Study Director*) is a senior program officer for the Board on Science Education at the National Academies of Sciences, Engineering, and Medicine. He has worked on a broad range of issues including informal science education, K-12 science education standards, federal science education programs, and K-12 engineering education. He recently held a 2-year position as a policy analyst in the Office of Science and Technology Policy, managing its Committee on Science, Technology, Engineering, and Mathematics Education, and providing the President and his senior staff with advice on science, technology, engineering, and mathematics education. Previously, he was a research associate at ICFi, an international consulting firm. He has an M.A. and a Ph.D. in applied developmental psychology from George Mason University.

S. JAMES GATES, Jr., is a university system regents professor, the John S. Toll professor of physics, and director of the Center for String and Particle Theory, all at the University of Maryland. He has made major contributions in the fields of supersymmetry, supergravity, and string theory. He is a member of the National Academy of Sciences, the American Academy of Arts and Sciences, and the American Philosophical Society. He is a fellow of the American Physics Society, the American Association for the Advancement of Science, the National Society of Black Physicists, and the British Institute of Physics. He serves on the U.S. President's Council of Advisors on Science and Technology and the Maryland Board of Education. He is also a member of the Board of Trustees of Society for Science & the Public and on the Board of Advisors for the U.S. Department of Energy's Fermi National Laboratory. He is a recipient of the Medal of Science, the highest recognition given by the U.S. government to scientists. He has a B.S. in mathematics, a B.S. in physics, and a Ph.D. in physics, all from the Massachusetts Institute of Technology.

SYLVIA HURTADO is professor and director of the Higher Education Research Institute at the Graduate School of Education and Information

Studies at the University of California at Los Angeles. Her primary research interests are in student educational outcomes, campus climates, college impact on student development, and diversity in higher education. She is past president of the Association for the Study of Higher Education. She has coordinated several national research projects, including a U.S. Department of Education-sponsored project on how colleges are preparing students to achieve the cognitive, social, and democratic skills to participate in a diverse democracy. She currently directs a national longitudinal study on the preparation of underrepresented students for science, technology, engineering, and mathematics careers. She has an A.B. in sociology from Princeton University, an Ed.M. from the Harvard Graduate School of Education, and a Ph.D. in education from the University of California, Los Angeles.

LEAH II. JAMIESON *(NAE)* is John A. Edwardson dean of engineering, Ransburg distinguished professor of electrical and computer engineering, and professor of engineering education, all at Purdue University. She is co-founder and past director of the Engineering Projects in Community Service Program. She is a member of the National Academy of Engineering and the American Academy of Arts and Sciences and a fellow of the American Society for Engineering Education and the Institute of Electrical and Electronics Engineers (IEEE). Jamieson is the current president of the IEEE Foundation. She is a recipient of the Bernard M. Gordon Prize for Innovation in Engineering and Technology Education from the National Academy of Engineering and of the Director's Award for Distinguished Teaching Scholars of the National Science Foundation. She has served as president and chief executive officer of the IEEE and board chair of the Anita Borg Institute. She has a B.S. in mathematics from the Massachusetts Institute of Technology and an M.A., an M.S.E., and a Ph.D. in electrical engineering and computer science, all from Princeton University.

ADRIANNA KEZAR is a professor for higher education at the University of Southern California and co-director of the university's Pullias Center for Higher Education. Her work focuses on change, governance, and leadership in higher education and her research agenda explores the change process in higher education institutions and the role of leadership in creating change. She is also a qualitative researcher and has written several texts and articles about ways to improve qualitative research in education. Her recent research projects include a study of networks formed to work with faculty in science, technology, engineering, and mathematics to improve undergraduate education. She has a B.A. from the University of California at Los Angeles and an M.A. and a Ph.D. in higher education administration from the University of Michigan.

KENNETH R. KOEDINGER is a professor of human computer interaction and psychology at Carnegie Mellon University. His research interests include creating educational technologies that dramatically increase student achievement. He is a cofounder of Carnegie Learning, Inc., and leads LearnLab, the Pittsburgh Science of Learning Center. He has created cognitive models, computer simulations of student thinking and learning that are used to guide the design of educational materials, practices, and technologies. With his colleagues, he has developed cognitive tutors for mathematics, science, and language and has tested them both in the laboratory and in classrooms. His research has contributed new principles and techniques for the design of educational software and has produced basic cognitive science research results on the nature of mathematical thinking and learning. He has a B.S. in mathematics, an M.S. in computer science from the University of Wisconsin–Madison, and a Ph.D. in cognitive psychology from Carnegie Mellon University.

JAY B. LABOV is the senior advisor for education and communication at the National Academies of Sciences, Engineering, and Medicine. He has directed or contributed to many institutional reports on undergraduate education, teacher education, advanced study for high school students, K-8 education, and international education. He oversees various activities for the institutions, including confronting challenges to teaching evolution and improving education in the life sciences. Previously, he was on the biology faculty at Colby College. He is a Kellogg national fellow, a fellow in education of the American Association for the Advancement of Science, and a Woodrow Wilson visiting fellow. He is a recipient of the Friend of Darwin award from the National Center for Science Education. He received a B.S. in biology from the University of Miami and M.S. in zoology and Ph.D. in biological sciences from the University of Rhode Island.

ELIZABETH O'HARE (*Program Officer*) is on the staff of the Board on Higher Education and the Workforce at the National Academies of Sciences, Engineering, and Medicine. Her portfolio includes projects that address science, technology, engineering, and mathematics workforce development, research administration and the higher education regulatory environment, and the competitiveness of American research universities and the scientific enterprise. Previously, she served as a legislative assistant for Representative Rush Holt (NJ-12), where she handled energy, science, and education policy issues. She began her work in science policy at the Society for Research in Child Development as a congressional science policy fellow of the American Association for the Advancement of Science. She has an A.B. in psychology from Bryn Mawr College and a Ph.D. in neuroscience from the University of California at Los Angeles.

MURIEL POSTON is the dean of faculty and vice president of academic affairs at Pitzer College. Previously, she was dean of the faculty and vice president of academic affairs at Skidmore College and a faculty member in the biology/botany department at Howard University. She has served as the division director for human resource development in the Education and Human Resources Directorate at the National Science Foundation (NSF) and as the deputy division director for NSF's Division of Biological Infrastructure. Her work has focused on supporting minority-serving institutions and underrepresented groups in STEM. Her research has been in the field of plant systematics, environmental law, and environmental policy. She is on the board of directors for the American Institute of Biological Sciences. She has a B.A. from Stanford University, an M.A. and a Ph.D. from the University of California at Los Angeles, and a J.D. from the University of Maryland.

MARK B. ROSENBERG is president of Florida International University, where one of his major efforts has been in STEM education, including partnerships with local schools, community colleges, and community organizations. He previously held the positions of chancellor and executive vice president for academic affairs at the university, as well as a faculty member. He previously also served as chancellor for the board of governors of the State University System of Florida. His research interests have been as a political scientist specializing in Latin America. He is a member of the Council on Foreign Relations and served as a consultant to the U.S. Department of State and the U.S. Agency for International Development. He has a B.A. in political science from Miami University and an M.A. and a Ph.D. in political science from the University of Pittsburgh.

HEIDI SCHWEINGRUBER is director of the Board on Science Education of the National Academy of Sciences, Engineering, and Medicine. She has served as study director for several projects, including the one that published *A Framework for K-12 Science Education*. She coauthored two award-winning books for practitioners that translate findings of NRC reports for a broader audience: *Ready, Set, Science! Putting Research to Work in K-8 Science Classrooms* and *Surrounded by Science*. Previously, she served as a senior research associate at the Institute of Education Sciences in the U.S. Department of Education and as director of research for the Rice University School Mathematics Project. She has a Ph.D. in psychology (developmental) and anthropology and a certificate in culture and cognition, both from the University of Michigan.

P. URI TREISMAN is professor of mathematics and of public affairs at the University of Texas at Austin and director of the university's Charles A.

Dana Center, which carries out policy research and evaluation in support of efforts to raise educational standards and to enhance the state's educational accountability system. His research focuses on designing programs that strengthen the teaching and learning of mathematics and science. He serves on the Policy and Priorities Committee of the Education Commission of the States and is a founding board member of the National Center for Public Policy in Higher Education. He also serves on the 21st Century Commission on the Future of Community Colleges, an initiative of the American Association of Community Colleges. He is the recipient of a MacArthur Fellowship for his work in developing programs to strengthen the education of minority and rural students. He has a B.S. in mathematics from the University of California at Los Angeles and an interdisciplinary Ph.D. from the University of California, Berkeley, where he studied both mathematics and education.

MICHELLE VAN NOY is associate director at the Rutgers University Education and Employment Research Center. Previously, she worked at the Community College Research Center at Teachers College at Columbia University, where she conducted studies of contextualized basic skills education, employer perceptions of the associate degree for information technology technician jobs, and community college noncredit workforce education. She also previously worked at Mathematica Policy Research, Inc. Her research focuses on the role of higher education, particularly community colleges, in workforce development. Her current work focuses on effective practices in community college workforce education, student decision making about majors and pathways through higher education, and linkages between education and employers. She has a B.A. in psychology and Spanish and an M.S. in public policy from Rutgers University and a Ph.D. in sociology and education from Columbia University.

X. BEN WU is a professor of ecology and presidential professor for teaching excellence at Texas A&M University. At the university, he formerly served as associate dean of faculties and director of the Center for Teaching Excellence with responsibility for organizing the efforts in faculty professional development in teaching. Previously, he was on the faculty of Ohio State University. His research interests are landscape ecology and undergraduate ecology education. One major interest is exploring technology-enhanced pedagogy for active learning, especially web-based and virtual authentic inquiry projects in large-enrollment introductory ecology classes and their effects on student learning. He has a B.S. in botany from the Lanzhou University and an M.S. in ecology and management science and a Ph.D. in ecology from the University of Tennessee.